ENERGY SECURITY AND CLIMATE CHANGE
A CANADIAN PRIMER

Edited by Cy Gonick

Fernwood Publishing · Halifax & Winnipeg

Canadian Dimension · Winnipeg

Editing: Glenn Bergen and Terry Murphy
Cover Design: Alexis Dirks
Printed and bound in Canada by Hignell Book Printing

Published in Canada by Fernwood Publishing
Site 2A, Box 5, 32 Oceanvista Lane
Black Point, Nova Scotia, B0J 1B0
and 324 Clare Avenue, Winnipeg, Manitoba, R3L 1S3
www.fernwoodpublishing.ca
and Canadian Dimension
2E-91 Albert Street, Winnipeg, Manitoba R3B 1G5
www.canadiandimension.com

Fernwood Publishing Company Limited gratefully acknowledges the financial support of the Government of Canada through the Book Publishing Industry Development Program (BPDIP), the Canada Council for the Arts and the Nova Scotia Department of Tourism and Culture for our publishing program.

Library and Archives Canada Cataloguing in Publication

Energy security and climate change: a Canadian primer / Cy Gonick, editor.

Includes bibliographical references.
ISBN 978-1-55266-248-9 (pbk.)

1. Climatic changes. 2. Petroleum reserves--Forecasting. 3. Petroleum industry and trade--Industrial capacity--Forecasting. 4. Power resources--Canada. I. Gonick, Cy, 1936-

QC981.8.C5C534 2007   363.738'74   C2007-903361-X

Credits

"Energy Futures" by Julian Darley is based on a presentation given on November 18, 2006, at the conference "Power for the People: Determining Our Energy Future," University of Alberta, Edmonton, Alberta.

"Responding to the Challenge of Peak Oil" by Richard Heinberg was originally published in *Canadian Dimension*, July/August 2006.

An earlier version of "Peak Oil and Alternative Energy" by Jack Santa Barbara was published in *Canadian Dimension*, July/August 2006.

An earlier version of "Scarring Tar and Scum from the Bottom of the Pit" by Petr Cizek was published in *Canadian Dimension*, July/August 2006.

An earlier version of "Power Speaks to Power under the Nuclear Revival Tent" by Marita Moll was published in *Canadian Dimension*, July/August 2006.

A shorter version of "Why Canada Needs a National Electricity Grid" by Marjorie Griffin Cohen was previously published as "U.S. is now determined to control Canada's electric power," ccpa *Monitor*, February 2006.

An earlier version of "Sustainable Agriculture—Has Cuba Shown the Way?" by Paul Phillips was published in *Canadian Dimension*, September/October 2006.

A longer version of "Bridging Peak Oil and Climate Change Activism" by Richard Heinberg was published in his book *Peak Everything: Waking Up to the Century of Declines* (New Society, 2007).

A longer version of "The Corporate Climate Coup" by David F. Noble originally appeared as a response to "Global Warming: Truth or Dare" by Dennis Rancourt, http://activistteacher. blogspot.com.

"Turning on Canada's Tap" by Tony Clarke was originally published in *Canadian Dimension*, September/October 2006. A longer version appeared in a chapter called "Turning on the Tap? Water Exports to the United States" in Bruce Campbell and Ed Finn, eds., *Living With Uncle: Canada-US Relations in an Age of Empire* (Lorimer, 2006).

"A Twelve-Step Program to Combat Climate Change" by Cy Gonick and Brendan Haley is a slightly expanded version of an article published in *Canadian Dimension*, March/April 2007.

Contents

PART 2: CLIMATE CHANGE / 107

Contributors

PETR CIZEK lived for over a dozen years in the Northwest Territories, working as a self-employed environmental consultant. He is currently a PhD student at the Collaborative for Advanced Landscape Planning, Faculty of Forestry, University of British Columbia. Working with the Dehcho First Nations, he independently designed and negotiated temporary protection for a 7 million hectare network of conservation areas in 2003, which was recognized as a "globally significant conservation" achievement through the Gift to the Earth Award from WWF International in Switzerland.

TONY CLARKE together with Maude Barlow, was awarded the 2005 Right Livelihood Award for his exemplary and longstanding worldwide work for trade justice and the recognition of the fundamental human right to water. Until his dismissal in 1993, Clarke was head of the social action department of the Canadian Conference of Catholic Bishops. In 1987 he chaired (first with Barlow, later by himself) the Action Canada Network, the largest coalition of civil society organizations and labour unions ever assembled in Canada to mobilize opposition to the free-trade agenda. He later founded the Polaris Institute "for the purpose of unmasking the corporate power that lies behind government."

MARJORIE GRIFFIN COHEN is an economist who is a professor of political science and women's studies at Simon Fraser University. Her most recent books are *Remapping Gender in the New Global Order* (2007), *Training the Excluded for Work: Access and Equity for Women, Immigrants, First Nations, Youth and People with Low Income* (2003) and *Governing Under Stress: Middle Powers and the Challenge of Globalization* (2004). Marjorie has served on several boards and commissions in British Columbia, including BC Hydro, the BC Power Exchange, the BC Industrial Inquiry Commission on the Fisheries and the BC Task Force on Bank Mergers. She was also instrumental in establishing the Canadian Centre for Policy Alternatives in British Columbia and was its first chair.

JULIAN DARLEY is founder and director of Post Carbon Institute. He is author of *High Noon for Natural Gas: the New Energy Crisis* (2004) and co-author of the forthcoming *Relocalize Now! Getting Ready for Climate Change and the End of Cheap Oil*. Julian has an MSc in environment and social research from the University of Surrey in the UK and an MA in journalism and communications from the University of Texas at Austin He currently lives in Sebastopol, California.

CY GONICK is founder, publisher and coordinating editor of *Canadian Dimension* magazine. He is executive producer of a weekly radio show called Alert. In 2000 he retired from the University of Manitoba after teaching there (political economy and labour studies) for over thirty-five years.

BRENDAN HALEY is energy coordinator at the Ecology Action Centre in Halifax. He is a member of the *Canadian Dimension* collective.

RICHARD HEINBERG is one of the world's foremost Peak Oil educators and is a research fellow of the Post Carbon Institute. He is the author of eight books, including *The Party's Over: Oil, War and the Fate of Industrial Societies* (New Society, 2003, 2005), *Powerdown: Options and Actions for a Post-Carbon World* (New Society, 2004) and *The Oil Depletion Protocol* (New Society, 2006). Richard is also a core faculty member of New College of California, where he teaches a program on culture, ecology and sustainable community.

GORDON LAXER is the director of Parkland Institute, an Alberta research network that studies public policy alternatives. He has been a political economist in the Department of Sociology at the University of Alberta since 1982. He is principal investigator for a five-year research project entitled "Neo-liberal Globalism and its Challengers: Reclaiming the

Commons in the Semi-Periphery: Canada, Mexico, Australia and Norway," principally funded by the Social Sciences and Humanities Research Council of Canada. Gordon is the author of *Open for Business: The Roots of Foreign Ownership in Canada*, for which he received the 1992 John Porter Award for the best book written about Canada, from the Canadian Association of Sociology and Anthropology.

DALE MARSHALL has worked as an environmental policy analyst for eight years, focussing exclusively on climate change for the last four. He has a master's degree in resource and environmental management from Simon Fraser University.

MARITA MOLL is a research associate with the Canadian Center for Policy Alternatives and a sessional lecturer at Carleton University in technology, science and environment studies.

DAVID F. NOBLE is a historian of technology, science and education. He is best known for his seminal work on the social history of automation, including *Forces of Production* (1975). He currently teaches in the Division of Social Science and the Department of Social and Political Thought at York University. His most recent book is *Beyond the Promised Land* (2005).

PAUL PHILLIPS is professor emeritus (economics) at the University of Manitoba where he taught for thirty-four years. He is also professor of American studies at the University of Ljubljana, Slovenia. He currently resides in Vernon, British Columbia, in semi-retirement, where he also writes a monthly column on environmental economics for the regional newsmagazine, *North of '50*.

JACK SANTA-BARBARA is a retired business executive who currently directs the Sustainable Scale Project (www.sustainablescale.org). His interests focus on how energy production and consumption drive ecological degradation, economic growth and violent conflict. He is involved with several local and international projects relating to sustainable communities, an ecologically and socially sustainable economy and a sustainable energy future.

KEVIN SMITH is a researcher with Carbon Trade Watch, a project of the Transnational Institute that studies the impacts of carbon trading on society and the environment

METTA SPENCER is Emeritus Professor of Sociology, University of Toronto, and sits on the executive of Science for Peace. She is the editor of *Peace* magazine. Her introductory textbook, *Foundations of Modern Sociology*, was published in nine editions. Another book, *Two Aspirins and a Comedy: How Television Can Enhance Health and Society*, was published in February 2006.

EDWARD SCHREYER served as Premier of Manitoba (1969 to 1977) and Governor General of Canada (1979 to 1984). He has had an interest in energy matters for many years and has taught on energy issues in university settings around the world.

JOHN W. WARNOCK has recently retired from teaching political economy and sociology at the University of Regina and is author of *Selling the Family Silver: Oil and Gas Royalties, Corporate Profits, and the Disregarded Public*, published by the Parkland Institute at the University of Alberta and the Canadian Centre for Policy Alternatives, Saskatchewan Division.

Foreword

by Rt. Hon. Edward Schreyer

It is remarkable when one comes across a work that is both profound and timely. Cy Gonick and the fifteen other authors who have contributed to this effort are to be commended for their insight, timely relevance and articulate explanation of the energy/environment context and times in which we live. This book is a treat. It is both primer and essential navigational chart for policymakers—which is to say for all of us who are concerned that we are drifting rudderless in terms of energy policy. Simply look at the pervasive evidence of denial and disconnect as each year passes with a continuation of the trends and trendlines of increased fossil-fuel consumption and consequential carbon dioxide emissions and associated climate change data.

It is reassuring to note that the contributed essays are cogent and candid in their identifying of both climate change and the fast-track depletion of oil and gas as the double threats to humankind's future existence and well-being.

What is needed is candor—in identifying the causes and probable mitigation. So too is candor needed in exposing the pervasive inconsistency which surrounds most policy seminars and conferences in the energy world. For example, have you sat through a presentation by one or another senior energy-sector executive who proceeds to describe how important it is to bring carbon dioxide emissions under some manageable limits along with developing complementary policies and efforts to develop renewable alternatives to oil and gas? I certainly have—and on many occasions now. Rather incredibly, however, in almost every case, when the industry spokesperson in the next presentation lays out their assumption as to future energy scenarios, it is one of linear incremental production and consumption of oil and gas and coal. Renewables of each and every kind are shown to increase, but at a rate that leaves their share of a total energy pie more or less constant to 2050. This is all errant nonsense! By 2050, at that rate, all will be beyond any remedial action. Thus far, far too few are prepared to tell the emperors (of the oil and gas interests) that they have no clothes.

But read the chapters in this book and you will be guided back to reality. You will gain some insight into the reality—of both American natural-gas supplies; of global oil peak (or plateau) scenarios; of the net energy input/output ratios inherent in ethanol. You will also read of some genuine concerns about the trading or sale of carbon emissions as opposed to any sustained, determined effort to reduce carbon dioxide by decarbonizing our energy sector and the economy generally by way of technology, conservation, alternative modalities, etc. The word *obscenity* has been used to describe this "easier way" of trading or even selling "rights" to emit carbon dioxide. As though it were a right to pollute—and as though it could ever be acceptable to sell such a right. You may wish to ponder this *Alice in Wonderland* logic. How many auditors will it take to ensure there is genuine capping and genuine offsetting? If it works, will it serve as a model for public health efforts to deal with global epidemic probabilities? That is to say, will we ever try to justify paying others to adopt preventive medical and health measures so that we can forgo any rigorous effort ourselves? Or perhaps, a more appropriate analogy is the one that is now exactly 500 years old and derived from medieval Europe: do we propose to sell indulgences so that we can continue to indulge in business as usual?

I urge you to read this book. You may come to some inconvenient conclusions. You will see why there is need for constant vigilance in our democracies in monitornng what is being said and then cross-checking with actual deeds.

Ed Schreyer
July 2007

Introduction

by Cy Gonick

If in the last century and a half humankind was overwrought by concerns over hunger, disease, war, poverty, imperialism, racism and economic depression—in this new century, the focus is shifting to climate change and energy security. Of course, climate change and energy security have not replaced these historical concerns; rather, they interact with and deepen them, making them more potent than ever. The US-led invasion and occupation of Iraq, for example, is clearly related to the increased importance of the Gulf region as the threat of peak oil grows near. The decision to invade Iraq, like the decision to invade Afghanistan and the decision to abort Hugo Chavez's Bolivarian revolution in Venezuela is entirely rational. These are actions to grab control of the world's energy supply and thereby exercise veto power over industrial and ideological rivals.

This collection of essays, mainly drawn from the pages of *Canadian Dimension* magazine, examines energy security and climate change separately. They *are* different, but they are also linked as Richard Heinberg's chapter "Bridging Peak Oil and Climate Change Activism" makes clear. He writes:

> Climate change has to do with carbon emissions and their effects—including the impacts on human societies from rising sea levels, widespread and prolonged droughts, habitat loss, extreme weather events, and so on. Peak oil, on the other hand, has to do with coming shortfalls in the supply of fuels on which society has become overwhelmingly dependent—leading certainly to higher prices for oil and its many products, and perhaps to massive economic disruption and more oil wars. Thus the first has more directly to do with the environment, the second with human society and its dependencies and vulnerabilities. At the most superficial level, we could say that climate change is an end-of-tailpipe problem, while peak oil is an into-fuel-tank problem.

As recently as twenty years ago, few of us were aware of either peak oil or climate change. It was back in 1956 that US geophysicist M. King Hubbert first revealed the findings of his research that oil production in the lower forty-eight states will have reached its peak between 1966 and 1972. His prediction was universally dismissed in the oil industry. What is more, even when it turned out that US oil production did peak in 1970, most regarded it as cyclical at best. Thirty-seven years later there are still some skeptics but by now the evidence is overwhelming that we are reaching global peak oil and global peak natural gas as well. The chapters by Richard Heinberg ("Responding to the Challenge of Peak Oil") and by Julian Darley ("Energy Futures") delve into this background and provide the most up-to-date information. Meanwhile, Paul Phillip's chapter "Peak Oil and the Future of Agriculture" explores the implications for agriculture.

As for climate change, it was only in June 1988 that James Hanson, director of the NASA Godard Institute for Space Studies presented evidence before the US Senate Energy and Natural Resources committee of global warming due to the emission of carbon dioxide and other greenhouse gases into the atmosphere. That same year the United Nations established the Inter-Governmental Panel on Climate Change (IPCC).

The new awareness of mounting ecological problems spawned the 1992 Earth Summit in Rio and a second Earth Summit on Sustainable Development a decade later in Johannesburg. The Kyoto Protocol, first negotiated in December 1997 and coming into effect February 2005, binds the 169 industrial nations that signed it to reduce their gashouse emissions to below 1990 levels by 2012. Industrialized countries were expected to reduce their emissions five percent below 1990 levels. Jean Chretien pledged 6 percent for Canada. This turns out to have been a pledge that few countries will realize except by allowing industrial polluters to buy pollution rights (called carbon offsets) from countries whose emissions dropped below 1990 levels. More on this below.

As it turned out, U.S. president George W. Bush pulled the US out of Kyoto in 2001 calling it too harmful to the economy. Canada, which had signed the accord but took few steps to reach its target under the federal Liberals, officially defaulted on its commitment when the Conservative party took over the reigns of government in 2006. The chapter by climate change analyst Dale Marshall provides a detailed account of the Harper government's approach to climate change from the time it assumed office.

Time Lines

In the absence of radical change, how much time do we have before energy shortages and global warming produce irreversible crises?

The scientific consensus is that a global warming of 2°C is the point at which we enter into likely probabilities of irreversible changes. At the same time, these changes could also occur before 2°C, and many, especially people in the global South, could experience impacts before 2°C.

Writing in July 2006, James Hansen maintains that the greatest immedi-

ate threat to humanity from climate change is associated with the destabiliza-
tion of the ice sheets in Greenland and Antarctica. A little more than 1°C, he
informs, separates the climate of today from the warmest interglacial periods
in the last half million years when the sea level was as much as sixteen feet
higher. Further increases in temperature by 2.8°C could lead to a rise in sea
level by as much as eighty feet, judging by what happened the last time the
earth's temperature rose this high three million years ago. "We have," Hansen
insists, "at most ten years—not ten years to decide upon action but ten years
to alter fundamentally the trajectory of greenhouse gas emissions" if we are
to prevent such disastrous outcomes from becoming inevitable.[1]

Julian Darley ("Energy Futures") believes that oil and natural gas combined will peak in 2010. Of the total petroleum consumed to date, he writes,
we have used 90 percent of our oil in the last forty-eight years and 50 percent
of it in just the last twenty odd years. "Anybody looking at this would say
'this isn't a sustainable system'—and it's not," he writes.

Founder and director of the Post Carbon Institute, Darley notes that the
production of conventional oil in Canada peaked in the early 1970s, not long
after the US. Since then the tar sands have been the source of an increasing
proportion of Canadian oil, most of it exported to the United States. But
tar-sands output is heavily dependent on natural gas which itself peaked in
late 2001 or 2002.

"Energy makes the world go around, quite literally," Darley reminds us.
Less energy, less economic activity. Or, as Richard Heinberg offers, "peak oil
could trigger the mother of all downturns." Petroleum is the backbone of our
transportation; in Canada natural gas heats 80 percent of our homes, is essential in the production of fertilizer, in plastics and many other synthetics.

It won't be many years before Canada will be unable to meet its own
internal consumption of gas—twenty years or less, according to Darley. But
NAFTA obliges Canada to maintain its level of exports to the US regardless.
"This is going to produce political turmoil," writes Darley, "because it's going to challenge NAFTA, challenge the way Canada does business, and it's
going to challenge the way people manage to not die of cold in their homes
in winter."

It is likely understood but needs to be emphasized that reaching peak
oil and gas does not mean that oil and gas production falls to zero. It means
that it is in decline, that consumption exceeds discoveries so that oil and gas
reserves are shrinking. Depletion analysts look to about a 2 percent per year
decline in oil extraction following the peak of global oil production, with
the rate increasing somewhat as time goes on. Regional natural gas decline
rates will be much steeper. As demand exceeds supply, prices will continue to
rise, affecting not only the cost of heating and transportation but the cost of
food and just about everything else. This is always an inconvenience for the

rich, but it will produce a growing crisis for working people and especially for the poor.

Peak oil is not the entire explanation of rising prices. The extraction of oil becomes more costly as the physical and chemical properties of oil fields deteriorate in the process of extraction. More water has to be pumped in to sustain the pressure for bringing oil to the surface and drilling is more expensive, especially in off-shore areas. Then there are the increased exploration costs for heavy oils, deepwater oil and gas and the massive costs of extracting oil from the tar sands.

In his contribution ("Capturing Revenues from Resource Extraction"), political economist John Warnock examines how Canadian governments tax excess profits (i.e., economic rents) in the mining industry compared to other oil producing countries around the world. He concludes that the comparatively minimal revenues extracted by Canada can be explained "because of the deep integration of the industry into the United States, the continental energy policy which is now entrenched in NAFTA, the presence of large transnational corporations that dominate the industry in Canada, and the fundamental government commitment to a neoliberal political economy since the early 1980s."

The issue, of course, is that royalties from resource extraction is a crucial source of funding investment in alternative energies and increasingly so as conventional energy sources become scarcer.

Alternative Energy Sources

Jack Santa-Barbara ("Peak Oil and Alternative Oil") surveys all of the known alternatives to conventional oil such as wind power, solar power, "clean oil" and biofuels like ethanol. Most renewable energy is of low intensity or low quality so that large amounts of energy must be invested in order to produce and deliver adequate supplies. Their net energy—the amount of usable energy left to do work after subtracting the energy used up to extract it and process it—is therefore low and in some instances like ethanol, barely positive.

The world's vast coal reserves are seen by some as a partial solution to peak oil. While coal is the dirtiest fossil fuel, it can be gasified to produce a high-quality diesel fuel. There are many issues here, not least the enormous cost of replacing conventional coal-based power plants with new ones. But the most serious aspect of "clean coal" is that it is energy intensive—greatly reducing the net energy of what is left. And the energy that goes into cleaning the coal will emit considerable amounts of greenhouse gases.

As geographer Petr Cizek points out ("Scouring Scum and Tar from the Bottom of the Pit"), the Alberta tar sands—today the source of half of

Canadian oil production, and by 2030 a projected three-quarters—consumes massive amounts of natural gas and water and destroys hundreds of thousands of square kilometers of boreal forest. Alberta has some of the highest per capita greenhouse gas (GHG) emissions in the world, three times higher than the Canadian average and six times greater than western Europe.

Marita Moll ("Power Speaks to Power under the Nuclear Revival Tent") challenges the view that nuclear power is a non-greenhouse gas emitting technology. Nuclear power is the source of half of Ontario's electrical energy needs and is being touted as a carbon-neutral replacement for natural gas to heat Alberta's tar sands. Even neighbouring Saskatchewan, the major supplier of uranium to nuclear sites around the world, is exploring a nuclear future for that province. Nuclear power is fraught with environmental dangers. While nuclear-power generation is carbon-neutral, refining uranium can emit a lot of carbon dioxide. There are no long-term, safe methods to dispose of spent uranium, and nuclear accidents can happen, with devastating results.

More generally, writes Richard Heinberg in his chapter " Bridging Peak Oil and Climate Change Activism," most strategies to keep the economy energized as oil and gas disappear imply increasing greenhouse gas emissions…. By now a disturbing trend becomes clear: the two problems of climate change and peak oil together are worse than either by itself."

"Turning on Canada's Tap," the chapter contributed by Polaris Institute director Tony Clarke is included in this collection because of climate change's expected impact in the central areas of the continent as warmer temperatures induce dramatic increases in evaporation, sucking rainfall back into the air and drying out lakes, reservoirs and fields. As well, with global warming leaving behind less snow for the rivers that rely upon them, rivers that serve 70 million people in a dozen states and four provinces, the demands for bulk water transactions within Canada and with the US could become overwhelming. As Clarke points out, a series of large-scale water diversion schemes for massive bulk water transfers from Canada to the US has been talked about going back four decades. For a variety of reasons, including overwhelming opposition by Canadians, none of these schemes were realized. But over the next few decades, as global warming takes place, the pressures from Washington to proceed will surely mount.

Responses to Climate Change

There have been three dominant responses to climate change. The first is to deny it or to consign its cause to natural conditions rather than to industrial systems and individual lifestyle practices. The second is to blame it mainly on individual lifestyle practices. The third is to insist that industrial and corporate behaviour can be changed through the market itself—taxing carbon emissions, thereby imposing a price on pollution; establishing pollution rights for industry that can be traded in a new carbon commodity market; and, more generally, providing an incentive system that rewards good environmental practices and punishes harmful ones.

While the media gives roughly equal weight to contrarian views, virtually none of the articles in peer-reviewed journals have questioned the consensus that emissions from human activities cause global warming. In his contribution to this collection, "The Corporate Climate Coup," historian David Noble traces the origins of climate change deniers and their corporate sponsors like Exxon Mobil, the world's largest single producer of greenhouse gases. Their intent, says Noble, has been to doubt, deride and dismiss "distressing scientific claims which might put a damper on enthusiasm for expansive capitalist enterprise."

Noble also traces a second corporate campaign that emerged somewhat later, in the wake of the growing awareness of global warming. Rather than denying it, this second corporate climate movement "sought to get out ahead of the environmental issue by affirming it only to hijack it and turn it to corporate advantage." It does this by locating "market-based" remedies quite compatible with corporate interests and practices—in this way turning away discussions of reform from more radical systemic solutions. This campaign has found a large following among economists and other professionals teaching and working for government, the media and non-governmental organizations. It fits their understanding of how the world works and how best to fix things that need repair.

The third response to climate change (and peak oil, for that matter) is to focus on what individuals can do to conserve energy and save the environment. In this volume Metta Spencer, in particular, gives emphasis to personal choices. But neither she nor our other contributors suggest that individualist strategies of sharing, repairing, reusing, recycling and cutting back can succeed in a society that screams out "BUY, CONSUME, WASTE!" and one that does not provide alternatives by radically restructuring our cities, living spaces and modes of transportation.

The perspective of our contributors is that both peak oil and climate change are real. But neither peak oil nor climate change arise primarily from wasteful or excessive personal lifestyle habits and neither can be managed by better gas mileage, voluntary cut-backs in consumption, walking and biking, carbon taxes, tradable pollution permits or incremental changes to business as usual. They are simply not proportional to analysis that points to the massive economic disruption that will result as energy supplies become more scarce and their prices soar; and to the growing threat to the planet, including species extinction, arising from global warming.

Kevin Smith ("Obscenity of Carbon Trading"), for example, takes on the most widely promoted market-based solution to climate change, namely carbon trading, which is enshrined in the Kyoto Protocol. As he writes:

> Effective action on climate change involves demanding, adopting and supporting policies that reduce emissions at source as opposed to offsetting or trading. Carbon trading isn't an effective response; emissions have to be reduced across the board without elaborate get-out clauses for the biggest polluters. There is an urgent need for stricter regulation, oversight, and penalties for polluters on community, local, national and international levels, as well as support for communities adversely impacted by climate change.

Smith points to *The Durban Declaration of Climate Justice*, signed by civil society organizations from all over the world, which asserts that making carbon a commodity represents a large-scale privatization of the Earth's carbon-cycling capacity, with the atmospheric pie having been carved-up and handed over to the biggest polluters.

George Monbiot, author of the best-selling *Heat, How to Stop the Planet From Burning*, compares the purchase of carbon offsets by polluters to the medieval practice in which the church sold indulgences to wealthy sinners. So long as you could pay the price you could go on sinning and still get to heaven!

Energy Security for Canada

In "The US Energy Act and Electricity: What it means for Canada" economist Marjorie Griffin Cohen asserts that unless the extension of recent US electricity policy into Canada is challenged it will force a radical and regressive change in the way Canadian hydroelectric power is produced and delivered, to whom it is sold and on what terms. To all intents and purposes this "regulatory imperialism," as she calls it, transfers control of the entire electricity system to US companies, integrating Canadian hydroelectric power into a continental grid and making it all but impossible to implement a national electrical grid which is essential for Canadian energy security.

In his chapter "Energy Security for Canadians" Parklands director Gordon Laxer describes the policies and practices that have led to the supply of Canadian energy being shaped to serve US energy security rather than Canadian and how this torpedoes any chance of addressing climate change in this country. Laxer writes:

> Until Canada grasps "energy independence," and a new "national energy policy," as US leaders call their plans, Canada cannot seriously address climate change. To do the latter, Canada would have

to regain control over energy exports, ownership and production levels. Canada needs a new energy strategy.

Laxer offers a multi-pronged program that, besides building a national electricity grid for power sharing across Canada, includes imposing an immediate moratorium on new tar sands projects; getting a Mexican exemption on compulsory energy exports to the US, or giving six months notice and exiting NAFTA; reinstituting the twenty-five years of "proven" supply of oil and gas for Canadians before export licences can be issued; insisting that 100 percent of the economic rents [excess profits] on existing oil and gas should go to the resource owners—citizens of energy-producing provinces and First Nations; and, following Canadian public opinion, nationalizing or provincializing part or all of the oil and gas industry.

Resource Wars

As global scarcity mounts and new industrial giants like China and India enter the race for energy and materials, the US is not only concerned to maintain its share of the world's energy supply—one out of every four barrels of oil of the world's production—but also to control the access of its rivals and potential rivals. This surely is the rationale behind the US invasion and occupation of Iraq. As Noam Chomsky has explained, "control [of the Middle East and its oil] gives the United States 'veto power' over its industrial rivals."[2]

China in particular is feared as an economic threat. Chomsky notes that the Bush administration's March 2006 National Security Strategy document describes China as the greatest long-term threat to US global dominance. It warns that Chinese leaders are not only "expanding trade but acting as if they can somehow lock up energy supplies around the world."[3]

China's oil imports are growing by 24 percent a year. To achieve energy security its state-owned companies have taken a stake in forty-four countries, prospecting for and producing oil and gas as well as constructing pipelines. Just before the US-led attack on Iraq, China had signed a production-sharing contract with Baghdad. It has developed close ties with Saudi Arabia, a US client state since the British were expelled during World War II, and with Iran, now China's third largest supplier after Saudi Arabia and Angola. These countries and others look to developing closer ties to China as a means of reducing their dependency on the US.[4]

Venezuela, the leading oil producer in the western hemisphere, is another concern, and China is investing there too as well as in other resource-rich Latin American exporters. Venezuela has been using its oil wealth to help the likes of Argentina, Bolivia and Brazil to pay off their debts to the International Monetary Fund, a financial tool Washington uses to control the economies of Third World nations. While Venezuela's Hugo Chavez has been nationalizing that country's oil industry, Bolivia's Evo Morales has been nationalizing that country's gas resources, the largest gas reserves in Latin America.

Russia, with 13 percent of the world's proven oil reserves and 34 percent of its gas reserves, is yet another major energy player. And together with China, India—Asia's other giant—has formed the Asia Energy Grid to ensure reliable delivery networks and energy security and to acquire stakes in production and exploration for which New Delhi and Beijing continue to cooperate and compete.[5]

"The days of unalloyed Anglo-American petroleum dominance are gone," writes William Tabb. He continues:

> This does not mean that Exxon-Mobil is not the world's most valuable and most profitable company nor that the oil giants do not benefit from high oil prices. They do, however, face more sophisticated national oil companies from China, India, Brazil and elsewhere who compete for supply that is increasingly under the control of state-controlled producers. The seven largest national oil companies, like Kuwait Petroleum, Abu Dhabi National Oil Company, Algeria's Sonatrach and the more familiar Saudi Aramco hold at least half the world's proven resources and account for a quarter of current production.[6]

As the case of Iraq demonstrates so clearly, the US, together with Britain, is using state terror and coercion to reassert its dominance of the world's energy system. Obviously it is a failing effort not only because of the strength of the Iraqi resistance, but because of the massive amount of energy the US must employ in conducting its foreign wars. Indeed, Michael Klare has calculated the annual tab in Iraq and Afghanistan alone at 1.3 billion gallons of oil, not counting the world's largest fleet of modern aircraft, helicopters, ships, tanks, armored vehicles and support systems maintained by the Pentagon.[7]

"[U.S.] 'oil security,'" writes the economist Elmar Altvater, "has several dimensions: first, the strategic control of oil territories; second, the strategic control of oil logistics (pipelines, the routes of oil tankers, refineries and storage facilities); third, influencing the price of oil by controlling supply and demand on markets; and fourth, determining the currency in which the price of oil in invoiced. When we consider the many strands in a complex strategy of oil security or "oil imperialism" the formula of "blood for oil" seems much too simple. Yet it is essentially correct."[8]

British defense secretary John Reid warns that climate change will further aggravate the resource wars. It will, he says, "make scarce resources, clean water, viable agricultural land even scarcer" and so "make the emergence of violent conflict more likely."[9] William Tabb adds that "in the United States, too, military planners and the CIA spin out scenarios of wars for desperately needed natural resources and the need to deal with the mass migrations of desperate people as entire societies disintegrate. Climate change, the forecasts suggest, will bring on new and even greater resource wars."[10]

Energy, Climate and Capitalism

In various ways both peak energy and climate change are tied to the very heart of the capitalist system. Capitalism is growth driven. Without economic growth profits cannot expand and without expanding profits capitalism would die. Besides a pool of available labour to keep wages in line, expanding production requires increased supplies of energy and materials. As conventional sources are exhausted, ever-resilient capitalism searches the earth and the oceans for new sources and invents new technology to overcome natural barriers. But over time these essential inputs become increasingly more costly to produce and deliver and undermine profit margins.

Besides fostering international conflict in the competition for and control over scarce resources, the scramble for resources inevitably comes into conflict with indigenous peoples on whose traditional lands they are most likely to be found and with local communities and environmentalists.

There are capitalist solutions to energy insecurity and climate change, but they mostly result in moving the problems around or causing other ones. For example, as conventional oil becomes increasingly scarce, the capitalist market system will go all out with coal, tar sands and nuclear power—exacerbating greenhouse gas emissions and imposing new levels of dangers with regard to nuclear accidents and nuclear waste. Further, the high cost of producing "clean coal," nuclear power and oil from the tar sands only adds to the increased cost of energy from conventional oil and gas. And as the market pushes up the price of energy in a world already marked by huge inequalities, the fall-out is a horrific worsening of conditions for most of the inhabitants of the world.

As the threat of climate change becomes ever clearer, around the world a new movement called eco-socialism is being born. The reason that this new movement was bound to appear and is bound to grow is that the capitalist system, however flexible and resilient it has been over some centuries of stress, has no acceptable answer to the twin crises of peak energy and climate change which it itself has produced. Setting hard limits on the use of carbon fuels whether by means of regulation or carbon taxes or any of the other tools discussed by experts in this volume interferes with economic growth, the sine qua non of capitalist accumulation.

In the words of the eco-socialist manifesto written by Joel Kovel and

Michael Lowy, this is an

> unacceptable option for a system predicated upon the rule: Grow or Die! And it cannot solve the crisis posed by terror and other forms of violent rebellion because to do so would mean abandoning the logic of empire, which would impose unacceptable limits on growth and the whole "way of life" sustained by empire. Its only remaining option is to resort to brutal force, thereby increasing alienation and sowing the seed of further terrorism ... and further counter-terrorism, evolving into a new malignant variation of fascism.

"In sum," the manifesto reads,

> the capitalist world system is historically bankrupt. It has become an empire unable to adopt, whose very gigantism exposes its underlying weakness. It is, in the language of ecology, profoundly unsustainable, and must be changed fundamentally, nay replaced, if there is to be a future worth living.[11]

There are many signs that the efforts to combat climate change are far too little, far too late. After all, carbon is a life-endangering issue. Yet, in this country even the very modest target set by the Kyoto Accord—that would only have cut economic growth rates by a few percentage points and was intended only as an interim first step—is regarded by both the Harper government and industry as too costly to profits. Another sign—when Stéphane Dion, now leader of the Liberal party, was asked in 2005 what he intended to do about the Alberta tar sands, the single biggest contributor to climate change, he responded: "There is no minister of the environment on earth who can stop this from going forward, because there is too much money in it." Now there is lots of talk about putting a price on carbon by way of a carbon tax, but only a very large tax would have a significant effect on business and consumer behaviour and of course it would be highly regressive. If it is set too low, it would not curtail emissions. If it were set sufficiently high, many corporations would shift their operations to countries without such taxes. In any event, there is no indication that governments are prepared to impose carbon taxes of any magnitude on citizens or businesses.

At this time most environmentalists seem ready to acknowledge that "business as usual" will not stop or even slow down climate change. But only at the fringes of the movement are they at a point of grasping that capitalism is *ecocidal*.

We need to build the eco-socialist movement!

We obviously cannot wait for a blueprint socialism to solve this problem for us. We need to act now. The final chapter of this book by Cy Gonick and Brendan Haley sets forth a "Twelve-Step Program" around which to rally a broad-based movement to combat climate change. It is not intended as the final word, but seeks instead to spark a discussion, promote electoral and civic action, and challenge industry and all levels of government.

To put it frankly, we need to fight for time. Pressing for these reforms like a moratorium on further tar sands development and others like a good-old fashioned increase in progressive income taxes and wealth taxes to fund the expansion of parks and community recreational facilities, public transit, health and education, social housing and retrofitting offers the best short-term strategy for decarbonization.

To get from the current impasse of a profit-driven energy-depleting capitalist world order to a new eco-socialist global society requires much hard work and much hard thinking. This book is one place to start for those interested in contributing to the conversation and joining this new global movement.

Notes

1. Jim Hansen, "The Threat to the Planet," *New York Review of Books*, July 13, 2006, pp. 12–16, cited by John Bellamy Foster, "The Ecology of Destruction," *Monthly Review*, February 2007.
2. Noam Chomsky, "Imminent Crises, Threats and Opportunities in the Middle East and Globally," *Monthly Review*, June 2007, p. 6.
3. Ibid, pp. 8–9.
4. Dilip Hiro, *Blood of the Earth* (Nation Books, 2007), Chapter 7.
5. Ibid, Chapter 6 and pp. 244–246. Siddharth Srivasta, "The Foundation for an Asian Oil and Gas Grid," *Asian Times Online*, December 1, 2005.
6. William K. Tabb, "Resource Wars," *Monthly Review*, January 2007, p. 38.
7. Michael T. Klare, "The Pentagon v. Peak Oil," TomDispatch, http://www.tomdispatch.com, June 17, 2007.
8. Elmar Altvater, "Fossil Capitalism," in Leo Panitch and Colin Leys, eds., *Coming to Terms with Nature*, (Monthly Review/Fernwood, 2007), p. 51.
9. Cited in William Tabb, op. cit., p. 36
10. Ibid., pp. 36–7.
11. There is not the space here to discuss what an eco-socialist might look like. In the eco-socialist manifesto, Kovel and Lowy give some hints: "Ecosocialism retains the emancipatory goals of first-epoch socialism, and rejects both the attenuated, reformist aims of social democracy and the productivist structures of the bureaucratic variations of socialism. It insists, rather, upon redefining both the path and the goal of socialist production in an ecological framework. It does so specifically in respect to the 'limits on growth' essential for the sustainability of society." The eco-socialist manifesto may be accessed from http://members.optushome.com.au/spainter/Ecosocialist.html. The principles of eco-socialism are further expanded upon in Michael Lowy, "Eco-socialism and Democratic Planning" in Panitch and Leys, eds., *Coming to Terms with Nature*.

PART 1

PEAK OIL, ENERGY
AND WATER SECURITY

Energy Futures

More Questions than Answers

by Julian Darley

The Energy Problem: The Peak and Decline of Fossil Fuels

The system I call kamikaze capitalism must have oil in order to go fast. The amount of total energy consumed by the world has gone up by almost a factor of three in approximately the last forty years. And a great deal of that increase is provided by oil and gas. The rate of coal increase was not nearly as quick, but coal is now the fastest rising source of primary energy, and has been for some years. Until very recently it was thought global coal reserves were measured in hundreds of years. New research is challenging that assumption.

Natural gas has been a star performer going up by nearly a factor of four in just forty years. It has certainly been one of the unseen drivers of the economy.

Gas and coal provide about half all the fossil-fuel energy we use (which is itself about 85 percent of our energy), but petroleum supplies not only the other half, but the half that will be hardest to substitute, because liquid fuels are so hard to make and are the backbone of the transport system. Most worryingly, of the total petroleum consumed to date, we've used 90 percent of our oil in the last forty-eight years and 50 percent of it in roughly just the last twenty. Anybody looking at this would say, "this isn't a sustainable system"—and it's not.

Energy makes the world go around, quite literally. Energy makes everything work, inanimate and animate—if anything moves, it's due to energy. Energy is also what makes the economy go, so if there's less total energy in the economy, I suspect we're going to see less economic activity.

Peak Oil

In 1956 M. King Hubbert predicted that oil extraction would peak in the US around about 1970, and so it did, and it has now declined to about half what it once was. Alaska's Prudhoe Bay was the larges-ever petroleum find in North America. It still couldn't pull extraction back to the peak, and extraction has fallen away precipitously. Now we are getting the same promises with the recent Jack 2 find in the Gulf of Mexico. But at 30,000 feet and under 7,000 feet of sea, this should be seen as a signal of total desperation. Not least because petroleum geology suggests that petroleum at that depth is a rare and unusual thing. The lesson of national peak oil is that you rarely ever return to the heights of production. You have to find another enormous oil province to get back to where you were.

The end of the age of oil discovery happened within a hundred or so years. Discovery peaked in 1964, but the years of the 1950s, 1960s and to some extent the 1970s were years of enormous discovery that in recent years has tailed off dramatically. The twentieth century was the age of oil discovery; the twenty-first century is going to be about something else. We are using up what we found in the last century, and we are using it up quickly.

Canada's Outlook: Oil
Canadian oil extraction peaked in the early 1970s, just after the US peaked, following which there was a drop and a very bumpy plateau. Now conventional oil appears to be in serious decline. Conventional oil—the stuff that's easy to get out and easy to refine—is what makes the economy go. Canada is one of the world's major oil producers and a considerable exporter. But Canada is also a huge importer—it's importing nearly a million barrels a day, and that number is going up quite starkly. Canadian imports have more than doubled in the space of about twenty years. And, in point of fact, these million barrels a day imported to the east coast are almost exactly equal to the whole of Canada's conventional production. So, without the tar sands and heavy oil, Canada would already be in some trouble and certainly wouldn't be able to export very much at all.

Canada's Outlook: Natural Gas and the Tar Sands
The tar sands depend very heavily on natural gas for all their extraction and upgrading procedures. A great deal of the revenues which make Alberta so rich actually come from natural gas, not from oil, and certainly not from the tar sands. There is an economic dependence on natural gas, and the tar sands themselves would more or less cease operating tomorrow morning if the gas spigot got turned off (which I am not suggesting would happen overnight). Gas is regional: it is hard to transport, especially across oceans, so you more or less have to rely on steel pipes, which means gas tends to be landlocked. It is very difficult and expensive to transport. For the most part, North America has to rely on its local supply.

Natural gas is fantastically convenient for industry, a wonderful process fuel and great for heating houses (about 80 percent of Canadian homes are heated with gas) and to make electricity with. It is also the main feedstock for nitrogen fertilizer, which is such a big part of industrial agriculture. It is used in the making of plastics and many other synthetics, including fabrics, and there are a vast number of uses for it in the petrochemical industry, which we are heavily dependent on, even if we aren't fully aware of them.

As already noted, the tar sands are also dependent on gas. The tar sands is about mining and melting. Mining takes slightly less than a thousand cubic feet to produce each barrel of synthetic crude, and the in-situ process takes much more. In situ will become the dominant way of extracting the bitumen, which does not flow. In order to make it flow, you have to melt it. Steam-assisted gravity drainage, a process that require huge amounts of natural gas, is the main way of doing this.

The supply situation is this: North America has about four percent of world reserves of natural gas, it is extracting about 29 percent and using 30 percent. The difference (1 per cent) comes in through liquefied natural gas (LNG). Canada itself has 1 percent of world reserves, but it is the world's third largest producer at the moment. Given all this demand for gas, this is simply not a situation that can carry on. And it is now reaching a critical point.

Production and discovery are very closely linked. Will future production go down as quickly as discovery? No one can be sure. I do not believe it will go down that fast. But I think we are clearly at a shoulder moment. And if it is a steep shoulder, North America is in terrible trouble. US natural-gas production has been declining at about 1.7 percent per year since 2001.

The number of US wells producing the gas is going up even as production is declining. This is a sure sign that a system of extraction is what is called "mature," but in my language, I would call it senile. This is very bad news if the goal is not only keeping a system going, but actually increasing it. And with the recent price falls in natural gas, drilling activity has started to tail off. I think we are going to see some additional trouble here—not only is the geological system itself showing signs of maxing out, but we are now drilling less in response to falling prices.

Canada, the US and NAFTA

From the point of view of the US, there would have been a gas crisis a long time ago had it not been for Canada. Canada has been exporting more than half its gas south for quite a long time, and the NAFTA treaty forces Canada to keep exporting the same proportion of the gas that it has been exporting on average for the last three years. So, of course, it would be a huge problem for Canada and for the US if Canadian gas were in trouble. And the supply of Canadian gas is in real trouble! The number of producing wells has nearly doubled in the last few years and yet Canadian gas peaked in late 2001 or early 2002, and it is in a bumpy decline. It has made some recovery, but with the decline in drilling we are going to see a serious fall in Canadian gas. Do not be deluded into thinking that production from the British Columbian part of the Western Canadian Sedimentary Basin or eastern Canada will help—they make up a tiny percentage of Canada's total production and they are in decline too.

One of these days, Canada is going to be challenged for its exports, and it won't even be able to meet its own internal consumption of gas. Gas extraction is going down, and the question is where will it meet domestic consumption? This is going to happen sometime in the next twenty years, probably sooner rather than later. This is going to produce political turmoil, because it will challenge NAFTA, challenge the way Canada does business and challenge the way people manage to not die of cold in their homes in winter.

Projections suggest a massive problem arising very soon. Eyes are turned to oil at the moment and to the tar sands. We depend very heavily natural gas, but we are not paying attention and I think people are going to be blindsided. North America has finally confessed that global warming is a problem, and we see that there might be some difficulties with oil, but we are relying on natural gas to come to our rescue as oil declines. This is certainly not going to happen in North America. Nor is it going to happen in Europe. Europe is right at the peak of its gas production, the UK is already in a precipitate fall. Only Norway has a great deal of spare capacity left. So, here we have the two great economic empires of the US and the EU all in gas decline.

The World Outlook

Now, with this situation, the idea is we will turn to the rest of the world and get more gas in that way, particularly via LNG—liquefied natural gas. But world natural gas discovery peaked in 1970 and has been in precipitate fall, as has oil. The world started using more oil than we discovered in 1981, and it has been getting worse more or less ever since then. The combined oil and gas picture shows a peak around about 2010.

Safe Carrying Capacity

We are in deep overshoot and need to move into what I call safe carrying capacity. Ecosystem limits are clearly shrinking, mainly because of our activity, so if we were to have anything like a sustainable species, we will have to move into a much smaller area, well within those ecosystem limits. William Catton in his extraordinary and powerful book *Overshoot* describes an environment's carrying capacity as its "maximum persistently supportable load." We need to understand carrying capacity, and how much more limited it is than we think. The Post Carbon Institute has developed a policy and a practice called *relocalization*, which I see as one of the only ways of getting within this safe carrying capacity.

The Response: Relocalization

Relocalization means reducing community consumption while aiming to meet daily needs through local production.

The Post Carbon Institute's Relocalization Network supports and connects over 150 local post-carbon groups around the world as they work toward relocalizing their communities. The Relocalization Network website (http://relocalize.net/) provides on-line communication tools to allow local groups to work together regionally and even globally (where appropriate) on local responses. The institute's forthcoming book, *Relocalize Now!* explores the theory and practice or relocalization. We have created an online repository of practical ideas, often contributed by local groups, that we've called *scripts*. These scripts can be added to and commented on the website.

Adoption of the Oil Depletion Protocol (http://oildepletionprotocol. org) is a way to reduce both consumption and production of oil on a grand scale. Originally proposed by Colin Campbell in 1996 and further developed by Richard Heinberg in a recent book (of the same name), the Oil Depletion Protocol is basically a global rationing system. Nations, citizens and organizations can adopt the protocol and work towards reducing their oil dependency by three percent a year, which would lead to a 25 percent reduction in ten years and a halving in roughly twenty years.

The institute has also published *Post Carbon Cities: Planning for Energy and Climate Uncertainty*, a guidebook for local governments responding to what is sometimes called "energy vulnerability." Municipalities and local government are going to be in the front line of energy and climate problems, so this is an important resource for decision makers and citizens alike.

The quickest way of reducing energy consumption is by sharing. For instance, car sharing is a very good way of getting transport oil consumption down very quickly in any industrialized country. Vancouver has an amazingly effective car-share organization as does San Francisco. Sharing directly challenges notions of individualism and it challenges notions of economic systems based on endless industrial growth. We are going to have to re-learn how to share, and car sharing is a good way to practise.

As we follow the unfortunate course of switching to industrial biofuels, we face the grave danger of grabbing more food land, more forest and more natural habitat just to turn it into biofuels. As a response, the Post Carbon Institute is spreading a network of local energy farms and gardens, first across North America and then hopefully beyond. The institute is experimenting with growing more food on less land, so we can make that spare land available for fuel, feedstock, fiber plant growing—and to give some back to forests. By forests, I don't just mean timber stands—I mean *forests*, the kind you leave alone for a thousand years. The greatest challenge is making these local energy farms productive with as close to zero petroleum product inputs as possible.

There is some good news and some bad news. The bad news is we are going to have to get our fingers dirty. The good news is we are going to have to get our fingers dirty. We are going to have to go back to making and maintaining things ourselves. When you get your fingers dirty and actually produce something useful and hopefully beautiful, it releases chemicals into your bloodstream, absolutely free and legally, and it makes you feel better. And you can't commoditize that.

City Design

I regard the grid model of city design as a complete disaster. Cities like Rome, Munich, Bratislava, Paris, San Sebastian, Copenhagen, Groningen and Venice were largely laid out before petroleum, often in a concentric model—very dense and compact in the middle, with the growing of food and other vital feedstocks around the edges. These places will work quite reasonably without petroleum: cars are largely excluded from them. I think it is true that making better cars will make worse cities. This is a very important lesson. Cars must be excluded from cities—it can be done, but it does take time and tremendous political will.

The design of cities is one of the dark secrets of North America: it has been largely laid out with petroleum in mind, and that is going to cause us huge problems. The only way we are going to get our petroleum consumption down seriously is by making some changes in how we have laid things out, and that is going to be very difficult. It takes decades to make these changes. We must start now.

There are some hopeful efforts. Sweden has pledged itself to be fossil free by 2020, and Oakland, California, is attempting something similar. Sweden has had a head start—it's already been doing it for thirty-five years, and a lot of its architecture and geometry is making that easier.

The note I wish to end on is this: we are going to have to get back to growing and making things locally and regionally. We shall need a local and regional renaissance. This renaissance—what we call relocalization—will rebuild friendship and society. It will transform everything.

Further Reading

Books:

Catton, William. *Overshoot* (University of Illinois Press, 1980).

Darley, Julian. *High Noon For Natural Gas: The New Energy Crisis* (Chelsea Green Publishing, 2004).

Darley, Julian, Celine Rich and Post Carbon Institute. *Relocalize Now! Getting Ready for Climate Change and the End of Cheap Oil* (forthcoming)

Lerch, Daniel. *Post Carbon Cities: Planning for Energy and Climate Uncertainty—A Guidebook for Local Governments* (Post Carbon Press, 2007).

Websites:

Global Public Media, http://globalpublicmedia.com. GPM has been featuring interviews with experts in oil and gas for several years.
The Relocalization Network, http://relocalize.net.
Oil Depletion Protocol, http://oildepletionprotocol.org.
Post Carbon Cities, http://postcarboncities.net.
Local Energy Farms Network, http://energyfarms.net.
Post Carbon Institute, http://postcarbon.org.

Responding to the Challenge of Peak Oil

by Richard Heinberg

Over the past two years the price of oil has climbed relentlessly. This is true not just for the volatile spot price, but also for the five-year futures price, which for many years held reliably close to the US$20 mark. At this time of writing, both the spot and futures prices are near $70. North Americans know that this translates to high gasoline prices, and, since they can scarcely live without their cars, many are worried and angry.

During these same two years, a small segment of the public has become aware of a phrase—"Peak Oil"—that both explains the recent price hikes and puts them in a broad historic context.

Those who "get" peak oil typically experience a profound paradigm shift—a reorientation of their thinking about the world. At the core of this shift is the recognition that humankind is entering a period of change unlike any in history. No one is saying that the world is about to run out of its most important strategic resource. However, the rate at which oil can be extracted is subject to geological limits. At some point those limits will begin to constrain our ability to produce oil at the ever-expanding rates that growing economies demand. What this means is that the amount of oil daily available to market will begin to decline, and the shortfalls will be relentless and cumulative.

There is of course some disagreement about when that point will be reached. Nevertheless, it is quite likely that the time interval before the global peak occurs will be much briefer than the period required for societies to adapt themselves painlessly to a different energy regime.

Not Just a Theory

What evidence is there that peak oil is real? A good part of the evidence comes from the fates of the older oil fields around the world. During the past century-and-a-half, all the older oil fields have been observed to peak in output and then decline. The same has been noted with entire producing nations. So it is axiomatic that, at some point, oil production for the world as a whole will also reach a maximum and then start to wane.

For large regions, such as nations, the rate of oil discovery typically peaks decades before the peak in actual oil production. This was the case, for example, in the US, which was the first important producing country to begin to decline. During the early twentieth century, the United States was the world's foremost producer and exporter of oil. Discoveries were dramatic and abundant, but after 1930, these began to fall off sharply. In 1970, the rate of US oil extraction reached its all-time maximum. And although later discoveries in Alaska and the Gulf of Mexico have offset this somewhat, the US rate has been in general decline since then. The result is that today the United States imports almost two-thirds of the oil it uses.

Today, according to ChevronTexaco, altogether thirty-three out of forty-eight significant oil-producing nations worldwide are experiencing declining production. In some cases, that decline may be temporarily reversible; however, in most instances it will continue inexorably.

Considering the importance of global peak oil, uncertainty regarding its timing is disturbing. If the peak were to occur within the next five years, it would be impossible for national economies to adjust quickly enough, while a peak thirty years from now would present a greater opportunity for preparation and adaptation.

Evidence for a near-term peak includes the fact that global rates of oil discovery have been falling since the early 1960s—as has been confirmed by Exxon. Currently, only about one barrel of oil is being discovered for every five extracted.

According to official data, current world petroleum reserves numbers look reassuring: the world has roughly a trillion barrels yet to produce. However, circumstantial evidence suggests that some of the largest producing nations may have inflated their reserves figures for political reasons. For example, a January 2006 article in *Petroleum Intelligence Weekly* discussed evidence that Kuwait's official reserves figures are double the amount that can actually be produced.

Uncertainties in the data invite disagreement among experts. While the US Department of Energy predicts that world oil production will increase over the next twenty years from 84 million barrels per day (Mb/d) to 120 Mb/d to meet anticipated demand, a growing chorus of petroleum geologists and other energy analysts warns that such levels of production may never be seen and that the global peak could occur within months or years.

Among those cautionary voices is that of James Schlesinger, who served as CIA director in the Nixon administration, defense secretary in the Nixon and Ford administrations and energy secretary in the Carter administration. In November 2005 testimony he offered before the Senate Foreign Relations Committee, Schlesinger urged lawmakers to begin preparing for declining oil supplies and increasing prices in the coming decades. "We are faced with the possibility of a major economic shock and the political unrest that would ensue," he said.

Ford Motor Company executive vice-president Mark Fields, in his keynote address to the Society of Automotive Engineers' 2005 Global Leadership Conference at the Greenbrier, noted that one of the most serious challenges to his industry is that "oil production is peaking."

Veteran petroleum geologist Henry Groppe, a Houston-based independent analyst who began his career in 1945 and who is today a consultant to global corporations as well as to nations, said in 2005 that "total crude oil production may have peaked this year, or perhaps will peak next year."

Matthew Simmons, founder of Simmons & Company International energy investment bank, has been perhaps the most outspoken of analysts regarding peak oil. He is the author of *Twilight in the Desert: The Coming Saudi Oil Shock and the World Economy*. Simmons has concluded, on the basis of his study of scores of technical papers from the Society of Petroleum Engineers, that Saudi Arabian oil production could be close it its maximum and that world oil production is also therefore close to its peak.

The Danger of Assuming an Easy Fix

These questions were addresses in an important study, "The Peaking of World Oil Production: Impacts, Mitigation and Risk Management," prepared by Science Applications International (SAIC) for the US Department of Energy under the leadership of Robert L. Hirsch in February 2005. The first paragraph of the report's executive summary states:

> The peaking of world oil production presents the US and the world with an unprecedented risk management problem. As peaking is approached, liquid fuel prices and price volatility will increase dramatically, and, without timely mitigation, the economic, social, and political costs will be unprecedented. Viable mitigation options exist on both the supply and demand sides, but to have substantial impact, they must be initiated more than a decade in advance of peaking.

As the Hirsch Report explains, due to dependence on oil for transportation, agriculture and the production of plastics and chemicals, every sector of society will be impacted. Since World War II, significant oil price increases have almost always led to economic recessions. Peak oil could therefore trigger the mother of all downturns—one that would be systemic and long-lasting. This lingering period of negative growth, widespread unemployment and skyrocketing real prices for just about everything would be the context within which nations would have to undertake heroic efforts to develop alternative sources of energy, reduce demand for oil through heightened energy efficiency and redesign entire systems (including cities) to operate with less petroleum.

The report effectively undermines the standard free-market argument that oil depletion poses no serious problem, now or later, because as oil becomes scarcer the price will rise until demand is reduced commensurate with supply. Higher prices presumably stimulate more exploration, the development of alternative fuels and more efficient use of remaining quantities. While it is true that rising prices will indeed do all of these things, we have no assurance that the effects will be sufficient, or will appear soon enough, to avert severe and protracted economic, social and political disruptions.

The upshot of the Hirsch Report is that if global peak oil is twenty years away or fewer, or we believe it might be, then we must begin immediately with a full-scale effort to address the problem.

The Oil Depletion Protocol

Peak oil mitigation efforts would be challenging enough in the context of a stable economic environment. But oil prices that repeatedly skyrocket and then plummet could devastate entire economies and discourage long-term planning and investment. Those nations, and those aspects of national economies, that could not obtain oil at any price they could afford would suffer the worst impacts. Supply interruptions would likely occur with greater frequency and for increasing lengths of time as global oil production gradually waned. Meanwhile the perception among importers that exporting nations were profiteering would foment animosities and an escalating likelihood of international conflict.

In short, the global peak in oil production is likely to lead to economic chaos and extreme geopolitical tensions, raising the spectres of war, terrorism and even famine, unless nations adopt some method of cooperatively reducing their reliance on oil.

One proposed method—perhaps the simplest imaginable—is the Oil Depletion Protocol, proposed by the Association for the Study of Peak Oil and the post Carbon Institute. According to its terms, signatory nations would reduce their oil production and oil imports according to a consistent, sensible formula that amounts to a bit less than 3 percent per year. This would have four principal effects:

First, it would reduce price volatility and enable nations, municipalities and industries to plan their economic future.

Second, it would reduce geopolitical competition for remaining oil supplies.

Third, it would conserve the resource. Petroleum engineers are keenly aware that oilfields that are depleted too quickly can be damaged, resulting in a reduction in the total amount eventually recoverable. Voluntarily reducing the rate at which the world's oilfields are depleted would extend their lifetimes.

And finally, carbon emissions would be reduced substantially—especially if the Oil Depletion Protocol were adopted in tandem with a strengthened version of the Kyoto Protocol. Oil consumption globally would decline over 25 percent in ten years, and this would translate into a sharp drop in greenhouse gas emissions.

The Oil Depletion Protocol would not by itself solve all of the problems raised by peak oil. But it should make those problems much easier to address, providing a context of global cooperation in which the task of energy transition could be planned and supported over the long term.

But could nations be persuaded voluntarily to reduce their consumption of a substance that has conferred so many economic advantages? Probably the strongest argument in this regard is the simple fact that oil is going to become scarce and expensive in any case: the Protocol simply enables nations to deal with the inevitable energy transition proactively.

One nation is already thinking along these lines. In December 2005, Swedish Prime Minister Göran Persson acknowledged that the global oil peak is a problem that needs to be addressed now, and announced the appointment of a National Commission on Oil Independence with the objective of making Sweden oil-independent by 2020. The commission will study mitigation measures and issue a report in the summer of 2007.

Municipal Peak Oil Planning

Some towns and cities around the world are unwilling to wait for their national governments to address the problem of peak oil and are looking for ways to prepare locally. Some examples:

Kinsale, Ireland, was the first town to undertake a comprehensive peak-oil assessment and response scenario, titled "The Kinsale Energy Descent Energy Action Plan." The project was initiated by Rob Hopkins and his practical sustainability class at Further Education College. The resultant report, with a year-to-year plan of action, has since been adopted as policy by the Kinsale town council.

In January 2006 the transportation committee of the City of Burnaby, British Columbia, released a report lucidly summarizing the challenge of peak oil and offering recommendations.

Sebastopol, California, has recently appointed a commission to study the problem of peak oil and make recommendations. I am proud to say that many of my students at New College of California are involved in the Powerdown Project, which offers assistance to Sebastopol and other regional towns in their efforts along these lines.

Other municipal, citizen-led efforts now under way include ones in Tompkins County, New York; the San Francisco Bay Area in California; Boulder, Colorado; Plymouth, New Hampshire; Bloomington, Indiana; and Eugene, Oregon, among many others.

A century from now humankind will no longer be using petroleum in any meaningful quantities. That much is assured. How we make the transition is up to us.

Further Reading

Flavin, C. and G. Gardner, "China, India and the New World Order," in World Watch Institute, *State of the World 2006* (Norton, 2006).

Heinberg, Richard, *The Party's Over* (New Society, 2005).

Heinberg, Richard, *The Oil Depletion Protocol: A Plan to Avert Oil Wars, Terrorism, and Economic Collapse* (New Society Publishers, 2006).

Hirsch, Robert L. et al., "The Peaking of World Oil Production: Impacts, Mitigation and Risk Management," prepared by Science Applications International for the US Department of Energy, 2005, http://www.projectcensored.org/newsflash/the_hirsch_report.pdf.

Simmons, Mathew, *Twilight in the Desert: The Coming Saudi Oil Shock and the World Economy* (Wiley, 2005).

Smil, Vaclav, *Energy at the Crossroads* (MIT Press, 2003).

Peak Oil and Alternative Energy

by Jack Santa-Barbara

Why Peak Oil Matters

The world is beginning to wake up to the fact that peak oil is real and coming to a neighborhood near you. Sweden and Norway have both initiated plans to be essentially free of fossil fuels by 2020, and a growing number of municipalities are beginning to incorporate energy consumption and production into their core planning activities. Various financial institutions have acknowledged peak oil, as well as some oil companies, independent geologists, the US Army Corp of Engineers and a range of corporations eager to cash in on alternative energy sources. While much of the world remains in a fog about peak oil and its implications, plans are already underway to prepare for an energy future that no longer relies on cheap energy.

Peak oil, of course, does not mean the world is about to run out of oil, but only that we are about to reach the level of maximum production of conventional oil—the half-way point, more or less, of our extraction of this unique resource. The implications of this geological fact are profound precisely because conventional oil is such a unique resource and because we rely so heavily on it to fuel not only our transportation services globally, but our economy as well.

The Uniqueness of Conventional Oil

What is so unique about conventional oil is that it has such a high energy content (compared to wood, for example), and is relatively easy to extract from the ground. Whereas the human energy that went into producing one BTU from wood was paid back some thirty times with the energy from burning the wood, we used to get a one hundred-times payback from conventional oil fields. The energy content of one barrel of oil is equivalent to manual labour from one person working for 12.5 years. In addition, oil is easily transported at ambient temperatures and relatively safe to handle. It also has many other uses besides producing energy—in plastics, pesticides and fertilizers, and in pharmaceuticals. With less oil, life will be very different than it is now.

For all its unique properties the feature of conventional oil that is most important is its high net energy—the amount of useable energy left to do work, after we subtract the energy we put in to extracting, processing and transporting it. And this is the crunch issue when we look for alternatives to conventional oil.

Conventional oil is a geologic deposit of fossil remains that is mixed with natural gas. Once a hole is drilled to reach it, the gas helps push the oil out of the ground to be collected and processed for use as a fuel. However, as a well becomes depleted, the gas is also diminished and no longer able to push the remaining oil out of the ground. The engineering solution to this phenomenon is to pump natural gas, carbon dioxide or water back into the ground to force out what remains. This works, but there is a cost. And the cost is not simply financial. It takes energy to pump water or gas into the well to extract more oil; as a consequence, the net energy (what is available to do work later) is reduced.

This reduction in net energy over time is a known characteristic of all conventional oil wells. Individual wells, oil fields and, indeed, entire national oil reserves demonstrate this phenomena. As the well or field matures, two things happen: the first is that the volume of oil that can be extracted from the source declines; the second is that the net energy also declines: it simply requires more energy to extract the remaining amount.

Approximately two-thirds of major oil producing nations are known to have peaked—that is, they have reached the half-way point of depleting their oil reserves. What is less well known is that as this peaking of individual national oil reserves occurs, the net energy available from these sources is also declining. The world ends up using more energy simply to extract more energy. The decline in net energy from oil can be measured: it has gone from 100:1 early in the twentieth century to less than 20:1 today; and it will continue to decline as more oil-producing nations, especially the large producers in the Middle East, reach their peak. When the global peak occurs, both the amount of oil extracted, and the net energy of that oil, will decline.

There is still debate about the timing of a global peak in oil production, but most credible estimates place the event within the next ten years. Uncertainty regarding the global peak is largely due to the secretive nature of the national oil industries (Saudi Arabia, Iran and Kuwait), whose officially stated reserves have enormous implications for their social, economic and political stability. There is an urgency to this issue which international bodies such as the International Energy Agency and the Council on Foreign Relations have only recently begun to refer to, albeit obliquely. It is questionable whether adequate time is available to prepare society for the transition that is about to occur.

A New Meaning to "TINA": There is
No Alternative to Conventional Oil

Maggie Thatcher's famous but incorrect quote that "there is no alternative" to capitalism can be more accurately applied to conventional oil. In terms of net energy, there is no alternative to conventional oil; and it is net energy that drives the global economy.

Coal and natural gas have net energy ratios almost as high as conventional oil. But coal is the dirtiest of the fossil fuels, and natural gas will peak not long after conventional oil peaks. Natural gas has already peaked in North America and now has to be shipped great distances at great costs. Such shipments are also dangerous as they are under high pressure and would cause enormous explosions if accidentally ignited or attacked by terrorists. As with the remaining easy oil supplies, the largest natural gas supplies are in geographic areas that are either politically unstable, or unfriendly to North American interests, or both. So these energy sources are problematic in terms of replacing the energy that will be lost when peak oil hits.

Non-conventional petroleum reserves (those in the deep ocean, the arctic and the tar sands) all require considerably more energy to extract and process than conventional oil. The huge reserves (said to be some 1 trillion barrels) in the Canadian tar sands have been touted as a potential solution to peak oil. But at best, oil from the tar sands will only create a small ripple in the overall impact of peak oil. The tar sands are projected to increase production to less than 3 million barrels a day by 2020, a very small part of the current global consumption of just over 80 million barrels a day.

The reason for this slow production from the tar sands is the enormous financial and energy costs involved. This oil is not drilled but mined; then the oil-impregnated sand is heated by natural gas for a week or more before it can be made to flow for processing. The net energy of the oil produced is much less than 5:1—significantly less than the ratio for conventional oil. In addition, there are considerable ecological and social problems involved in this mining operation, including the diversion and pollution of enormous quantities of water.

The Energy Ratios for the Main Alternatives

When we look at non-petroleum based energy sources, we find that they all have lower net energy ratios than conventional oil. Although in near ideal conditions wind can be as high as 30:1, in more normal conditions, wind has an energy ratio of approximately 18:1 and hydropower has a ratio of approximately 12:1. There are many challenges associated with wind turbines; there are long-term storage issues and electrical grids generally require a base load, currently provided by coal or nuclear generating plants. Many of these challenges are being addressed, but it will be some time before all the bugs are worked out. While hydroelectric power can have a moderate net energy ratio, most of the sites suitable for such projects have already been used, so we should not expect a significant amount of energy to come from this source in the future.

Geothermal power can be used to generate electricity at approximately 8:1, but there are few sites worldwide where the heat from the earth is close enough to the surface to make this source feasible.

Nuclear energy is now receiving more attention after years of neglect following Chernobyl and Three Mile Island. However, the net energy from various nuclear generating plants is only somewhere between 10:1 and 3:1. The problems with nuclear energy are well known: costs, radioactive waste, terrorist targets and they rely on non-renewable resources. In addition, a lot of fossil fuels would be used in both building the reactors and mining the uranium, so they are hardly "climate friendly."

Energy from the sun can be used directly to heat air or water, or by converting it to electricity. Direct use of solar energy was known to the Greeks and Romans who designed their buildings and cities to make optimal use of the sun's heat. Capturing direct solar energy using simple materials is known as passive solar, and it can have a low net energy (approximately 5:1). Accompanied by mirrors and or lens, it can be used to cook food, heat water to a boil and even melt certain metals. While useful, passive solar technologies are limited in what the energy provided can do.

Solar energy is more versatile by converting it into electricity. This requires somewhat more sophisticated technologies such as photovoltaic cells. Recent innovations have increased the efficiency of PV cells to over 40 percent. However, these innovations have not yet been brought to market, nor has their net energy been calculated; nor have life cycle analyses been performed to assess their overall environmental impacts. Nonetheless, these active solar technologies can be expected to generate net energy ratios in the 15:1 to 10:1 range.

The net energy from hydrogen depends on how the hydrogen is produced. Hydrogen itself is not an energy source, but a storage medium for energy. On its own, hydrogen cannot do any work. It requires energy inputs, generally from electricity, to then generate electricity. But all currently known methods of using hydrogen as part of an energy system get considerably less energy out than what is put in. This generates a negative net energy ratio. So if there were a more direct way of using the input energy (e.g., electricity), it would be more efficient to use the electricity directly.

Ethanol is also of questionable value from a net energy perspective. Various studies debate whether the net energy from ethanol is even positive. Ethanol from corn is very energy intensive and even when the equations look positive, they are only slightly positive, bringing into question whether corn-based ethanol makes any sense in energy terms. While ethanol burns cleaner than fossil fuels, where is the environmental benefit if a roughly equal amount of fossil fuels are used to make the ethanol?

Ethanol from sugar cane as developed in Brazil has a higher net energy (about 8:1 or 7:1), because it is less energy intensive to grow sugar cane than corn. But the extensive use of cane-based ethanol in Brazil is causing massive social upheaval, as indigenous people are displaced to expand the industrial cane fields.

It has to be kept in mind that we currently rely on fossil fuels for over 80 percent of our primary energy needs. Replacing the current infrastructure that makes use of these fossil fuels will itself require enormous amounts of fossil-fuel energy—all at a time when both the amount and net energy of fossil fuels will be declining. In addition, there remain technical challenges with many of the renewable energy technologies which have to be resolved before they can be scaled up to begin replacing fossil fuels.

It is also important to keep in mind that there is yet no standardized method of calculating net energy ratios. Some calculations are more comprehensive in terms of the energy inputs included than others. Consequently, net energy ratios can be influenced by those doing the calculations to obtain the results they desire. Differences in net energy ratios for the same energy source can therefore be either a function of somewhat different technologies or subject to influence from special interests.

Environmental and Social Implications of Alternative Energy

The fossil fuels that have created the enormous financial wealth evident in the world today have been a disaster in terms of climate change. The emissions of greenhouse gases have upset the carbon balance to the extent that global ecosystems are being altered, perhaps irrevocably. With hurricanes Katrina and Rita in the Gulf of Mexico in 2005, the melting of glaciers and tundra, and the possible extinction of the polar bear due to global warming we are beginning to get a glimpse of what the benign term *climate change* will mean. The climate chaos ahead will only worsen as we emit more greenhouse gases.

But peak oil is not likely to be a boon for reversing the climate trend for several reasons. First, there is still a lot of both conventional and non-conventional oil available, and none of the producing nations seems prepared to leave their liquid gold in the ground. Furthermore, coal is also abundant, and we are seeing a resurgence of coal processing both to produce liquid fuel and generate electricity. China is building dozens of dirty, old-style coal plants.

The technology of converting coal into a gas or liquid has been known since the Nazis sought to ensure their fuel supply in World War II by learning to liquefy coal. These processes are now being touted as the answer to peak oil. So-called "clean coal" technologies are also available which actually do remove an enormous amount of the impurities from coal. However, even the small amount of impurities that remain would contribute to local and regional pollution. In addition, the cleaning process does not remove the carbon emissions. Mining coal and processing coal is a water intensive process, and the quality of coal (in terms of both impurities and energy density) is waning. Much of the remaining coal is in wilderness areas, and would disrupt habitat and destroy biodiversity. Mercury from coal emissions will continue even after the coal is cleaned.

But the most serious aspect of clean coal is that it is energy intensive—greatly reducing the net energy of what is left. Dirty coal provides a net energy range of 80:1 to 60:1. Cleaning the coal can reduce this ration to less than 10:1. And the energy that goes into cleaning the coal will emit considerable amounts of greenhouse gases. One proposal for clean coal indicates that if this plan was followed the level of greenhouse gases by the end of this century (500 ppm) would be higher than the "high-risk" level described by the UN Intergovernmental Panel on Climate Change.

Part of the clean coal and coal liquefaction routines call for capturing and sequestering the carbon emission produced, recognizing the need to keep them out of the atmosphere. And it is true that there is some experience with pumping carbon dioxide into the ground. But what we do not know is whether it will stay there. Whatever leakages occur will continue to force climate change, and the *bequestration* risk will remain for many decades.

Biofuels may play a role in a sustainable energy future, but their low net energy is only one of their drawbacks. Agricultural land for fuel is already competing in some areas with land for food. As petroleum-based fertilizers are less available after peak oil, crop yields could decline, requiring more land to feed the same number of people. An increasing population will present even greater challenges.

But even renewable energy sources, such as solar and wind, can be used unsustainably. The more energy we produce the more matter gets moved; and the more matter moved the greater the impact on ecosystems. We need a new index of efficiency, one that minimizes the energy inputs per unit of human well-being.

In addition to these, and other, environmental costs of various energy alternatives, there is also the issue of equitable access to energy to meet basic human needs. While we in North America consume at least twice the amount of energy we need for a high level of well being, close to 2 billion people do not have access to electricity. Such inequities are not sustainable.

Net Energy and Net Profit

Will the environmental and social concerns of various energy alternatives be considered in how the world deals with peak oil? It seems unlikely, for there are enormous profits to be made in developing alternatives to conventional oil.

One of the fundamental mistakes being made by policy makers at very high levels is their unchallenged assumption that the energy loss with peak oil has to be replaced. Energy makes the global economy grow, and without energy it will decline. Most leaders see this as unthinkable. From an ecological perspective it is a likely necessity.

The Organisation for Economic Co-operation and Development (OECD) and World Trade Organization (WTO) are discussing a multinational, multi-trillion dollar investment in clean coal and nuclear technologies over the next several decades. The WTO through the General Agreement on Trade in Services (GATS) is currently attempting to privatize what are now state decisions regarding several critical aspects of energy services which would seriously impede governments' capacity to set energy policies.

Many governments around the world, including the Canadian federal and several provincial governments, are investing in and supporting ethanol production. Many are taking a new look at nuclear. Almost daily there are reports of President Bush touring an alternative-energy operation, touting its contribution to energy independence. Energy is seen as both an opportunity for profit and as a necessity for maintaining economic growth. Environmental and social sustainability are hardly mentioned. Indeed, environmental protection and previously supported environmental treaties such as Kyoto are being scrapped. Market forces, supported with the appropriate lobbing efforts, are being relied on to bring forth the best alternative energy sources. The problem, of course, is that the market only defines "best" in terms of financial profit.

A Framework for a Sustainable Energy Future

What the world needs now even more than energy is a framework in which to think about energy. The most basic questions of such a framework might be: what do we need energy for? How much energy do we need for a comfortable level of human well-being?

How do we ensure that scaling up energy alternatives will not push any critical ecosystems beyond a tipping point where those ecosystems no longer provide the services we rely on for our well-being? How do we set a priority to ensure all of humanity has access to their basic energy needs?

The current policy framework places economic growth as the top priority and recognizes the significance of energy to maintain such growth. This framework includes assumptions that such growth is required for human well-being and that any negative externalities (environmental or social) will be costed into the alternative energy systems that are brought to market. Such a framework is fatally flawed and is likely to destroy more ecosystems, create greater inequities and likely also generate more violent conflict. It will also make some people very rich, which is the reason why this is the policy trajectory we are now following.

An alternative policy framework would focus on the use of energy to meet basic human needs and differentiate clearly between economic growth and well-being. There are many reasons why economic growth per se is not only an inadequate indicator of human well-being but is actually a dangerous and misleading indicator. There are too many "negatives" in terms of environmental and social well-being that are considered as "positives" from a simple economic perspective (e.g., more cancer cases, food banks, refugee camps, auto accidents, etc. all contribute to Gross Domestic Product but clearly do not contribute to quality-of-life). What are needed are new quality-of-life indicators that take a broader perspective of what human well-being and happiness are genuinely about.

A Canadian contribution to this issue is the Genuine Progress Indicator (GPI), which includes not only economic indicators but social and environmental ones as well. When looked at from this broader perspective, one can see that social and environmental indicators can decline even as economic ones increase. This calls into question the economists' assumption that continuously increasing economic growth leads to better quality-of-life.

There is considerable evidence that subjective ("happiness") measures of well-being are only related to income at very low levels of income (from zero to about $10,000 in purchasing power parity—a index that takes into account the value of a dollar in different countries). Beyond this level of income there is no consistent increase in happiness.

A similar picture emerges if we look at objective levels of well-being (e.g., infant survival, female longevity, access to nutritional food and educational opportunities) and per capita energy consumption. Energy is extremely important for well-being but only up to a point. A fairly decent level of well-being is achievable with between 50 and 70 giga joules per capita per year; and optimal levels are achieved with about 110 GJ per capita per year. Beyond that there is no further increase on these objective measures. Interestingly, there is no relation between levels of energy consumption and the Human Freedom Index; high levels of freedom (clearly a quality-of-life issue) can be achieved with low levels of energy consumption.

Despite the optimal per capita level of energy consumption being in the 110 GJ range, we in North America annually consume about 325 GJ per person; the average Afghani consumes about 20 GJ. Social justice suggests that the over-consumers like us could reduce our energy consumption (without loss of well-being) and assist those at the bottom of the energy pyramid to meet their basic needs for energy.

These facts have enormous significance for a sustainable energy future. If our civilization continues to pursue economic growth as the means of achieving human well-being, we will likely wreck the planet and make inequalities worse. If we reorient our policies toward meeting fundamental human needs and remaining within the biophysical limits of the ecosystems that provide all our material wealth, perhaps we have a chance. Whether we can make this significant paradigm shift will be the key issue for human civilization as the multiple challenges of climate change, peak oil and massive extinctions present unprecedented threats not only to our well-being but even our survival.

All of us need to pay considerably more attention to the cumulative environmental impact of our collective energy production and consumption and ensure that environmental tipping points are not exceeded. Frugality needs to become one of the highest virtues if the several billion people on this planet are to approach a sustainable equilibrium with the biosphere that contains and sustains us.

Further Reading

Anielski, Mark. "The Genuine Progress Indicator: A Principled Approach to Economics," http://www.pembina.org/pdf/publications/gpi_economics.pdf.

Brown, Leter. "Beyond the Oil Peak." *Plan B 2.0.* 2006. http://www.earth-policy.org/. See Chapter 2.

Campbell, C.J. and Jean Laherrere. "The End of Cheap Oil." *Scientific American,* March 1998.

Council on Foreign Relations. "Independent Task Force Report #58—National Security Consequences of Oil Dependency." 2006.

Daly, Herman. "Economics in a Full World." *Scientific American,* September 2005.

Daly, Herman and Josh Farley, *Ecological Economics: Principles and Applications.* Washington, DC: Island Press. 2004.

Hirsch, R., R. Bezdek and R. Wendling, "Peaking of World Oil Production: Impacts, Mitigation & Risk Management." US Dept of Energy requested assessment, February 2005.

Hubbert, M. King. Testimony. Hearing on the National Energy Conservation Policy Act of 1974, before the Subcommittee on the Environment of the Committee on Interior and Insular Affairs, US House of Representatives. June 6, 1974.

International Energy Agency. "World Energy Outlook 2006." http://www.worldenergyoutlook.org/.

Menotti, Victor. "The Other War: Halliburton's Agenda at the WTO, A Policy Brief on the Energy Services Negotiations in the World Trade Organization (WTO), International Forum on Globalization, June 2006." http://www.ifg.org/reports/WTO-energy-services.htm.

"Peak Oil and the Environment," Sustainable Energy Forum 2006. Washington, DC. May 7–9, 2006. Audio recordings of conference proceedings available at http://www.sef.umd.edu.

Rees, William. "A Sunshine Limit to Growth." *The Ecologist* 20, 1 (2002), Energy Bulletin.

Smil, Vaclav. *Energy at the Crossroads.* Boston: MIT Press, 2003.

Sustainable Scale Project. http://www.sustainablescale.org.

Tainter, Joseph et al. "Resource Transitions and Energy Gain: Contexts of Organization." *Conservation Ecology* 7, 3 (2003): 4.

Scouring Scum and Tar from the Bottom of the Pit

Junkies Desperately Seeking One Last Giant Oil Fix in Canada's Boreal Forest

by Petr Cizek

Faced with the undeniable reality of "Hubbert's Peak" in global conventional oil supplies, the world's largest multi-national energy corporations are now hell-bent on squeezing oil out of tar in northern Alberta. They are like junkies desperately conniving for one last giant fix in a futile attempt to quench America's insatiable "addiction to oil," so eloquently described by President George W. Bush. Along the Athabasca River near Fort McMurray, a sub-arctic town almost 1000 kilometres north of the US border, tar literally seeps out of the riverbanks where Aboriginal peoples once used it to patch their birch-bark canoes. But most of the tar sands lie hidden below northern Alberta's boreal forest in an area larger than the state of Florida.[1]

The first serious effort to dig huge tar pits along the Athabasca river and steam out the oil started in 1963 with the Great Canadian Oil Sands Company developed by Sun Oil Ltd., later to become Sunoco and eventually Suncor. By 1967, the company's scion J. Howard Pew, self-proclaimed "champion of free enterprise and enemy of godless communism," had sunk $240 million (over $1 billion in today's currency) into this project in an effort to wean North Americans from dependence on foreign oil.[2] However, this Pew family project was less than successful, since separating the tar from the sand and then turning it into crude oil requires huge amounts of energy, steam and water. Even after the tar is melted down into bitumen, it still has to be "upgraded" into synthetic crude oil by adding hydrogen, usually made from natural gas.[3]

Starting in the 1970s, the federal and Alberta governments provided billions of dollars in tax breaks and research subsidies, invested in a joint-venture corporation called Syncrude, and even significantly lowered the royalty rate in the tar sands in 1997. In the mid 1980s, new "in-situ" technologies such as steam-assisted gravity drainage (SAGD) were developed, which steamed the tar from deposits around Cold Lake and Peace River that were too far below the surface to strip mine. By the turn of the millennium, dozens of multinationals had invested over $24 billion into the tar sands, which finally began to yield huge profits with the explosive increase in the price of oil.[4]

Between 1995 and 2004, tar sands production doubled to more than 1.1 million barrels per day, sixteen years ahead of schedule. It was a banner year in 2003, when the United States Energy Information Administration recognized that 175 billion of the total 1.7 trillion barrels of oil in the tar sands were "economically recoverable," placing Alberta second only to Saudi Arabia in the world's pecking order of oil reserves.[5]

The Tar Sands' Appalling Impact

"Oil Sands Fever," a report released in November 2005 by the Pembina Institute, a Calgary-based energy think-tank, barrages the readers with an onslaught of horrifying facts and images where all the appalling environmental impacts of the tar sands assume gargantuan proportions:[6]

- About half of Canada's oil production currently comes from the tar sands. Oil production from the tar sands is predicted to quintuple from 1 million barrels per day to 5 million barrels per day between 2003 and 2030, representing over three-quarters of Canada's oil production, 70 percent of which is destined for export to the United States.
- Around Fort McMurray, over 430 square kilometers of boreal forest has been eradicated, there is an approved disturbance of 950 square kilometers, and a planned disturbance of 2000 square kilometers. Not including the loss and fragmentation of boreal forest from "in-situ" operations in Cold Lake and Peace River, this will be twice the combined urban footprint of Calgary and Edmonton at 1000 square kilometers. No land has yet been certified as reclaimed.
- To produce one barrel of oil, four tonnes of material is mined, two to five barrels of water are used to extract the bitumen, and enough gas to heat 1.5 homes for a day is required. Oil sands producers move enough overburden and oil sands every two days to fill Toronto's Skydome or New York's Yankee Stadium.
- The tar sands industry now consumes 0.6 billion cubic feet of natural gas per day, enough natural gas to heat 3.2 million Canadian homes for a day. By 2012, they will consume 2 billion cubic feet of natural gas per day, enough to heat all Canadian homes for a day. By 2030, the tar sands are forecast to consume over 5 billion cubic feet per day of natural gas, representing more than the combined output of the planned Mackenzie Valley and Alaska gas pipelines, which will induce exploration and development of thousands of natural gas wells and feeder pipelines spanning the Northwest Territories, Yukon and Alaska.

- The greenhouse gas intensity of tar sands production is almost triple that of conventional oil, largely due to the vast amounts of natural gas consumed. Even before the actual produced oil is burned, carbon emissions from the tar sands are forecast to increase from 23.3 million tonnes per year to between 83 and 175 million tonnes per year. This might represent almost two-thirds of Canada's 2005 "Kyoto Gap" of 270 million tonnes. Canada's "Kyoto Gap" has increased from 138 million tonnes in 1997 to 270 million tonnes in 2005, largely due to the impact of the Alberta tar sands.

- Approved oil sands mining operations are already licensed to divert 349 million cubic metres of water per year from the Athabasca River. This is approximately three times the volume of water required to meet the municipal needs of Calgary, a city of almost 1 million people, for a year. Planned projects will increase water diversions to almost 500 million cubic metres of water per year, representing almost half of the river's winter low flow.[7]

- Syncrude and Suncor are the top two air polluters in Alberta. They have already degraded the once-pristine air quality in Fort McMurray, a small northern city of 71,000, to that of metropolitan centers like Edmonton and Calgary with populations of close to 1 million each. Air quality modelling for approved projects predicts that national, provincial and international guidelines for sulphur dioxide and nitrogen oxide will all be exceeded.

- Very recent developments reinforce and amplify all these facts. Multinational corporations, including majority players such as ExxonMobil, ConocoPhillips and Shell who also happen to be the promoters of the Mackenzie Valley and/or Alaska arctic natural gas pipelines, have confirmed a grand total planned investment of $100 billion into the tar sands in the next decade.[8] This makes it the largest mega-project complex in the world, growing to an astounding total of at least $160 billion when all the required arctic natural gas supplies from the Mackenzie Valley and Alaska pipelines are added, which requires coining "gigaproject" as a whole new word.[9]

A Modern Gold Rush

Brokers such as Raymond James Ltd. and Canadian Imperial Bank of Commerce are advising investors about the realities of peak oil and counselling them to invest in the tar sands, which they describe as the planet's last new significant oil supply addition with a total output that will rival Saudi Arabia by 2030.[10] When interviewed by CBS's *60 Minutes* in January 2006 about the confirmed tar-sands reserves of 175 billion barrels, Clive Mather,

CEO of Shell Canada, baldly stated that "we know there's much, much more there. The total estimates could be two trillion or even higher"—eight times the reserves of Saudi Arabia.[11]

Early in 2006, frenzied speculation hit new heights in a bidding war for new tar sands leases, where energy corporations spent almost twice as much cash in one month than they had ever been spent in a whole year. During the month of January 2006, the Alberta government raised $850 million for selling 4000 square kilometres of new tar sands leases. A grand total of 24,000 square kilometres of boreal forest is now available for extracting oil from tar, a combined area almost as big as Vancouver Island.[12] Finally, the Alberta government was steamrolling towards a March 31, 2006, deadline to complete public consultations about its Mineable Oil Sands Strategy (MOSS). The plan is to effectively re-zone the entire Fort McMurray region as a permanent industrial landscape. This would finally abandon the fantasy of multiple-use through "integrated resource management" that balances resource extraction with wildlife and other ecological values.[13]

The option of nuclear power to supply steam and hydrogen to the tar sands has been promoted for at least a decade by the federal crown corporation Atomic Energy Canada Ltd., desperate to rejuvenate the dwindling market for its white elephant CANDU reactors.[14] In the fall of 2005, the French energy conglomerate Total SA began to muse publicly about using nuclear power in its new tar-sands projects.[15] By January 2007, Husky Oil, majority owned by Hong Kong billionaire Li Ka-shing, added its voice to the expression of nuclear interest in the tar sands.[16]

To capitalize on these emerging opportunities, the Energy Alberta Corporation was formed in 2006, with an exclusive franchise from Atomic Energy Canada Ltd. to market CANDU reactors to the tar sands.[17] At the same time, the federal minister of natural resources Gary Lunn, responsible for Atomic Energy Canada Ltd., has become an enthusiastic promoter of nuclear power for the tar sands, partly due to nuclear's alleged benefits of reducing greenhouse-gas emissions.[18] More pragmatically, TransCanada Pipelines announced serious discussions to revive the dormant 1800 million watt Slave River hydroelectric project at the massive rapids near Fort Smith at the Northwest Territories border, host to the northernmost rookery of the once-endangered white pelicans.[19]

Our Meek Environmentalists

Despite Pembina's exhaustive documentation of this unfolding ecological holocaust, its policy prescriptions were shockingly timid and meek. On December 1, 2005, led by the Pembina Institute, eleven major environmental organizations, including World Wildlife Fund Canada, Canadian Parks and

Wilderness Society and the Sierra Club of Canada issued a declaration called "Managing Oil Sands Development for the Long Term." The declaration calls for elimination of subsidies, a network of protected areas and corridors, a binding integrated regional resource management plan and most curiously "carbon neutral (zero net greenhouse gas emissions) by 2020 through a combination of on-site emission reductions and genuine emissions offsets." The possibility of slowing down, much less stopping, further expansion of the tar sands was not even mentioned in the declaration.[20]

Moreover, the declaration went on to assure its intended audience that:

> These conditions can be implemented without significant macroeconomic impacts through innovation and strong leadership. Only through the satisfaction of these conditions do we believe that Canada will be in a position to develop this energy resource in a responsible manner that will create a positive legacy for current and future generations.

Half a year after this declaration was released, the Pembina Institute's Dan Woynillowicz seemed to have second thoughts and obliquely told the *Washington Post*, "We shouldn't be issuing new permits. We are foreclosing our future. In the 1990s, we acknowledged environmental challenges would occur. But we are seventeen years ahead of schedule."[21]

However, in early summer 2006, former Alberta Conservative Premier Peter Lougheed described the situation around Fort McMurray as "a mess" and publicly demanded a moratorium on further tar-sands expansion.[22] Heeding the lead of this well-respected arch-conservative, the Pembina Institute and the Canadian Parks and Wilderness Society finally called for a moratorium themselves in their report about "in-situ" steam-assisted gravity drainage (SAGD) "Death by a Thousand Cuts" released on August 1, 2006.[23] Thus far, the World Wildlife Fund has remained silent on the moratorium issue.

Are these environmental organizations so terrified of upsetting this "foundational economic driver for Canada"? These are the words used to describe the tar sands in a 2004 agreement between Alberta and the federal government that allows speedy work permits for foreign "guest-workers" to relieve labour shortages in Fort McMurray.[24] Or are there deeper forces yanking at these environmentalists' chains?

The Pembina Institute just happens to make money selling "carbon offsets" to the guilt-ridden rich and "progressive" corporations by subsidizing TransAlta Utilities, a major Alberta coal producer, to construct windmills that produce electricity without emitting carbon. Without legislated "caps" on maximum carbon emissions, the purchase of carbon offsets is a feel-good illusion, as this does not actually result in the reduction of overall carbon emissions. A major purchaser of these credits is Suncor, who most recently offset 238,000 tonnes of carbon out of its total of over 11 million tonnes per year of steadily growing carbon emissions—a whopping 2 percent.[25] The Pembina Institute also offers franchise opportunities to other non-profit groups in this money-making scheme.[26] With newly elected Conservative Prime Minister Stephen Harper's announcement that Canada cannot meet its Kyoto commitments, Pembina's climate-change specialist Matthew Bramley conceded with a capitulation: "We do need to buy international credits as part of our package of measures to meet our Kyoto target."[27]

Twisting Science

One need look no further than the list of "partners and clients" in the Pembina Institute's annual reports[28] to find a veritable who's who from the oil patch. The list includes major players in the tar sands/arctic natural gas pipelines "gigaproject" such as ConocoPhillips, Encana, Husky Oil, PetroCanada, Shell Canada and even Suncor, whose vice-president of sustainability makes valuable use of his surfeit of uncritical face time in Pembina's tar-sands video to proselytize about the company line.[29]

In November 2005, the Pembina Institute also produced a report called *Counting Canada's Natural Capital: Assessing the Value of Canada's Boreal Ecosystems*, wherein its lead author, ecological economist Mark Anielski, proclaimed that the Canadian boreal forest absorbs 173 million tonnes of carbon worth $1.85 billion each year, which could presumably be used to "offset" carbon emissions from the tar sands.[30] Yet, deep within the body of the text, Pembina acknowledged recent scientific literature, which concluded that the boreal forest actually became a net *source* of 44 million tonnes of carbon emissions per year in the 1970s, largely due to increases in forest fires and pest outbreaks, all related to global warming.[31]

In an amazing twist of logic, the Pembina Institute then used an earlier study[32] that explored possible intensive management scenarios, such as tree-planting, fire suppression and pest control to reverse future carbon emissions from the boreal forest as a way to rationalize its deceptive conclusion that the boreal forest is now absorbing carbon and not actually producing it. The primary author of all these boreal carbon studies is Dr. Werner Kurz, who despite Pembina's distortion of his studies, is surprisingly also listed as an advisor, reviewer and contributor to the report.

In January 2007, the same authors released another "natural capital" report focusing on the Mackenzie River watershed called the "Real Wealth of the Mackenzie Region."[33] Despite privately admitting that his earlier statement about the annual value of carbon absorption was inaccurate,

Anielski instead broadened his distortions. This time, he simply conflated carbon storage (stocks) with annual carbon absorption (flows): "The stored carbon and annual carbon absorbed by forests, peatlands, wetlands and tundra are valued at an estimated $252 billion in 2005, or 56 percent of the total estimated non-market value of ecosystem services." Even the *Globe and Mail* was hoodwinked by this remarkable sleight of hand and unquestioningly reported Pembina's earlier figure of $1.9 billion as the annual value of carbon absorbed in the Mackenzie region.[34]

Follow the Money

The sponsor of both these "natural capital" reports is the Canadian Boreal Initiative (CBI), self-described as "an independent organization working with conservationists, First Nations, industry and others to link science, policy and conservation activities in Canada's boreal region."[35] This "organization" does not have a board of directors, is not registered as a non-profit corporation under any federal or provincial laws and does not have a charitable number. Yet it employs a staff of eleven and has sponsored at least six major research reports about the boreal forest in the past three years. Former federal climate change negotiator, Cathy Wilkinson, the "director" of the CBI appeared on Report on Business TV's show "SqueezePlay" on December 13, 2005, to publicize the findings of the "Counting Canada's Natural Capital" report:[36]

> Kevin O'Leary: "I like to understand where people's interest lies and I always do that by finding out where their money came from because I believe that money and interests are always aligned. Who funds your organization?"

> Cathy Wilkinson: "We're funded by charitable organizations both in the United States and in Canada who are interested in the value of the world's largest and intact forests. So this is really an international effort in the same way that our trade moves across the border."

Not quite. As noted on CBI's own website, its sole funder is actually the Philadelphia-based Pew Charitable Trusts—the same Pew family who originally developed the tar sands, created Suncor (worth $5.9 billion when they sold their shares in 1995) and who continue to own Sunoco, a major refiner of synthetic crude oil from the tar sands.[37] So, how does CBI actually channel money "across the border" if it is neither a legal entity nor a registered charity in Canada? The Pew Charitable Trusts first transfers money to the North American headquarters of Ducks Unlimited in Nashville Tennessee, which is funnelled to its Canadian branch-plant headquarters in Winnipeg. Ducks Unlimited, who proudly lists Suncor and Syncrude as its "corporate partners," then recycles "boreal" money not only back to the CBI, but also to other Canadian environmental organizations such as World Wildlife Fund (WWF) and the Canadian Parks and Wilderness Society (CPAWS).[38]

Buying the Environmental Movement

WWF and CPAWS got their first snort of this nose candy from the Pew Charitable Trusts in 1999 with a grant of $1.8 million over two years to "protect at least 20 million acres of boreal forest wilderness in the Yukon and Northwest Territories of Canada." Even though they never came even close to this lofty objective, the Pew Charitable Trusts kept cranking the dose of crackerjack cash higher and higher from $2.1 million per year in 2000, to $4.5 million per year in 2002/2003, topping out with a mind-blowing speedball injection of $12 million—the Pew's single biggest grant in 2004.[39]

WWF and CPAWS also do not oppose the Mackenzie Valley pipeline in the Northwest Territories. After all, the new president and CEO of WWF, Mike Russill, is a former senior executive of Suncor. WWF and CPAWS are only lobbying to establish a network of protected areas prior to the start of construction. In 2004, WWF, CPAWS and Ducks Unlimited squeezed out an additional $9 million from the federal government to conduct more studies to identify such a network under the government-approved Northwest Territories Protected Areas Strategy.[40]

In 1996, WWF threatened to sue the federal government for allowing Canada's first diamond mine despite the absence of any protected areas in the central arctic region. WWF then withdrew its lawsuit in exchange for the government's commitment to develop a Northwest Territories Protected Areas Strategy. Since this strategy was approved in 1999, WWF has not formally identified, much less established, any protected areas in the central arctic, which has since been almost completely staked for diamonds and more recently uranium. After almost a decade of "implementing" the strategy in partnership with government, only two small peninsulas on Great Bear Lake have been permanently protected.[41]

In Alberta, the once-feisty Edmonton branch of CPAWS is pursuing a similar approach around the tar sands where it is promoting four protected areas, all of which have low oil potential and no leases.[42] This pattern of environmental organizations adopting a docile "low-hanging fruit" strategy soon after being bankrolled by the Pew Charitable Trusts has been thoroughly documented by American activists and investigative reporters.[43]

Setting Up Fronts

Until the early 1980s, the Pew Charitable Trusts religiously followed J. Howard Pew's founding mission "to acquaint Americans with the evils of bureaucracy, the paralyzing effects of government controls on the lives and activities of people, and the values of the free market" by funding right-wing extremists like the John Birch Society and the Heritage Foundation.[44] Since then, the Pew Charitable Trusts has adopted a much more sophisticated strategy of setting up dozens of socially progressive "front-groups" across North America like the CBI, an innovative right-wing twist on classic Marxist-Leninist organizing tactics.

In 2003, the CBI established the Boreal Leadership Council composed of WWF, CPAWS, Ducks Unlimited, Suncor, Tembec, Alpac, Domtar and several First Nations. Its goal is to implement the Boreal Conservation Framework by protecting at least half of Canada's boreal forest, with the remainder available for "leading-edge sustainable practices."[45] Concerned that this arbitrary benchmark of protecting half the boreal forest has never been scientifically justified, the David Suzuki Foundation did not endorse the framework. As pointed out by free-market environmentalist Larry Solomon of Energy Probe, the Boreal Conservation Framework actually amounts to a massive resource giveaway requiring government subsidies as industrial development in the far northern boreal forest is not economically viable under current market conditions.[46]

Lacking any accountable governance structures and directed by carefully chosen Pew operatives, front-groups such as the CBI not only fund, but also insist on entering into partnerships with established advocacy organizations. Thus, the front-group can serve as a "drag-anchor" on any activities that are excessively disruptive to the status quo. According to an employee of CPAWS, the CBI is now reviewing and vetting their draft press releases. A project officer of a major Canadian foundation recounts tales of environmental organizations pleading with him for grants, desperate to break their dependence on the CBI/Pew as their sole funding source.

The Sierra Club of Canada originally took a hard-line position against the Mackenzie Valley pipeline and the tar sands. In a wacky turn of events shortly after receiving funding from the CBI/Pew, Sierra Club endorsed the Pembina Institute-orchestrated "oil sands declaration" and threw its support behind the Alaska pipeline as the "lesser of two evils" on the entirely false assumption that none of the Alaska gas would go to the tar sands.[47] Sierra's flip-flopping continues with its most recent announcements that it now wants a moratorium on further oil-sands development[48] but is willing to accept the Mackenzie Valley pipeline as long as none of the gas goes to the tar sands.[49]

At the Boreal Forest Forum, CBI's annual secret conclave held at a rural retreat center outside of Ottawa in fall 2006, the Pembina Institute's presentation acknowledged how far back environmental organizations had lagged behind political and public opinion on the issue of a tar-sands moratorium.[50]

Pembina then stressed the need to "catch up and get ahead" on the moratorium issue, but also to "balance oil sands 'shock and awe' with solutions" such as protected areas, land use and pre-tenure planning, and best practices. We do indeed live in strange times, when mainstream environmental organizations lag far behind public opinion and even conspire to drag it down.

The only bright light in this deep dark pit of depravity is the report *Fueling Fortress America* released in March 2006 by the Parkland Institute, Canadian Centre for Policy Alternatives and the Polaris Institute. This report clearly advocates a moratorium on further tar-sands development, a national energy policy and exemption like Mexico from the "proportional sharing" clause of the NAFTA free-trade agreement, which only allows Canada to reduce energy exports to the United States in proportion to reductions of our own consumption.[51]

"Canadian Oil is Ours"

While allegedly supporting boreal conservation in Canada, the Pew Charitable Trusts is also busy addressing American foreign policy regarding the tar sands. With its partner, the $6 billion Hewlett Foundation, which gave the Pembina Institute $400,000 and funnelled another $1.85 million through the Pew Charitable Trusts to the cbi in 2004,[52] the Pew Charitable Trusts established the National Energy Commission, a bipartisan group of twenty energy experts. This group includes James Woolsey, a former director of the CIA and a member of the Project for a New American Century that advocated pre-emptive war and the invasion of Iraq as early as 1998. In its 2004 report, "Ending the Energy Stalemate," the commission advocated "a $300 million increase in federal funding over ten years to improve the environmental performance of technologies and practices used to produce unconventional oil resources" in Alberta's tar sands and Venezuela's Corinoco heavy-oil belt. Overall, the commission advocates $36 billion in government subsidies to the energy sector to be financed though the sale of carbon credits.[53]

Canada's tar sands and arctic natural gas have been on America's foreign policy radar screen since at least 2001, when Dick Cheney's National Energy Policy report stated that the continued development of the tar sands "can be a pillar of sustained North American energy and economic security."[54] Last year, American politicians were outraged by a relatively small Chinese investment of $225 million in the tar sands, which prompted energy analyst Irving Mintzer to blurt out the widely held but publicly unspeakable opinion of Beltway insiders: "The problem with the Chinese is that they don't know that the Canadian oil is ours. And neither do the Canadians."[55] Mintzer is also a co-author of the report *U.S. Energy Scenarios for the 21st Century* commissioned by the Pew Center on Global Climate Change.[56]

In 2005, Dick Cheney's National Energy Policy was transmuted into legislation called the *Energy Policy Act* with the support of the Set America Free Coalition, a front-group of the Institute for the Analysis of Global Security. A member and advisor to both organizations, former CIA Director James Woolsey proudly proclaims that "We've got a coalition of tree-huggers, do-gooders, sodbusters, hawks, and evangelicals."[57] This coalition includes the Natural Resources Defence Council, a regular recipient of millions in Pew and Hewlett funding, which also has become involved in a boreal forest campaign in northern Manitoba concerning hydroelectric exports to the mid-western United States.

In addition to providing $13.6 billion in subsidies to the energy sector, the Energy Policy Act has a "Set America Free" sub-section, which establishes "a United States commission to make recommendations for a coordinated and comprehensive North American energy policy that will achieve energy self-sufficiency by 2025 within the three contiguous North American nation area of Canada, Mexico, and the United States." Since the United States currently imports 60 percent of its oil supply, this continental self-sufficiency will require much additional Canadian and Mexican oil. Furthermore, the Act establishes a task force to initiate "a partnership with the Province of Alberta, Canada, for purposes of sharing information relating to the development and production of oil from tar sands." Finally, the Act requires the secretary of energy to update its assessment of domestic heavy oil resources to include "all of North America and cover all unconventional oil, including heavy oil, tar sands (oil sands), and oil shale."58 However, this legislation merely formalizes the continental "deep integration" that has been well underway with the establishment of the Security and Prosperity Partnership, which held its Oil Sands Experts Working Group in January 2006.59

Beware of Foundation Grants

Canadian environmental organizations should think long and hard about accepting money from and becoming financially hooked on the largesse of gigantic American foundations. Although there is no recent evidence of such activities, the CIA has a long history of funneling money through philanthropic foundations to achieve American foreign policy objectives by co-opting the soft non-radical Left. This is carefully documented in Francis Stonor Saunders' book *The Cultural Cold War*. In 1976, a select committee of Congress found that CIA funding was involved in nearly half the grants made by 163 foundations in the field of international activities.[60]

While countries like Sweden and Iceland are seriously planning to break free from oil by 2020, multinational corporations are busy digging Canada deeper into a tar-pit, determined to become history's last kingpin pushers to oil junkies across America and around the world, no matter the consequences to the planet's climate. The double tragedy is that so much potential opposition to this self-destructive lunacy has been readily defused through a wad of cash, an addiction just as cunning, baffling and powerful as oil. Members and individual donors to environmental organizations need to hold their leaders and staff to a much higher standard, making sure that they match their rhetoric with results, instead of just bouncing from one mega-project cash cow to the next in an endless hustle to line their pockets with lucre.

Notes

1. Alberta Department of Energy, *Oil Sands—Statistics for Public Offerings*, 2006, http://www.energy.gov.ab.ca/183.asp.
2. G. Laird, *Power: Journeys Across an Energy Nation* (Toronto: Penguin, 2002).
3. D. Woynillowicz, D. Severson-Baker and M. Raynolds, *Oil Sands Fever: The Environmental Implications of Canada's Oil Sands Rush* (Calgary: Pembina Institute for Appropriate Development, 2005), http://pubs.pembina.org/reports/OilSands72.pdf.
4. Laird, *Power*.
5. Alberta Department of Energy, *Oil Sands*.
6. Woynillowicz et al., *Oil Sands Fever*.
7. D.W. Schindler and W.F. Donahue, "An impending water crisis in Canada's western prairie provinces," in *Proceedings of the National Academy of Sciences* 103, 19 (2006): 7210-16, http://www.pnas.org/cgi/content/abstract/103/19/7210.
8. Government of Alberta, Statistics and Publications, *Inventory of Major Alberta Construction Projects—Oil Sands*, 2006, http://www.alberta-canada.com/statpub/albertaConstruction-Projects/mpgetem.cfm.
9. I. Austen, "Tricky Years of Maneuver Ahead for Proposed Gas Pipeline," *New York Times*, February 23, 2006, http://www.nytimes.com/2006/02/23/business/worldbusiness/23pipeline.html?ei=5070&en=cf1dfe8e366106ae&ex=1178769600&pagewanted=print.
10. J. Rubin and P. Buchanan, *The Time of the Sands* (Toronto: CIBC World Markets Inc., 2006), http://research.cibcwm.com/economic_public/download/occ_rep56.pdf; R. Mawdsley, J. Mikhareva and J. Tennison, *The Oil Sands of Canada: The World Wakes Up First to Peak Oil, Second to the Oil Sands of Canada, After Almost a Century of Research and Development, the Oil Sands Time Has Come* (Calgary: Raymond James Equity Research Ltd., 2005), http://www.raymondjames.ca/rjl_marketing/Library/Static/Assets/Documents/PDF/MKT/OIL_SAN_report0705.pdf.
11. CBS News, "The Oil Sands of Alberta," January 20, 2006, http://www.cbsnews.com/stories/2006/01/20/60minutes/printable1225184.shtml.
12. Alberta Department of Energy, "Oil Sands," 2006, http://www.energy.gov.ab.ca/89.asp.
13. Alberta Department of Energy, "About Mineable Oil Sands Strategy (MOSS)," 2006, http://www.energy.gov.ab.ca/3009.asp.
14. J. Donnely and D. Pendergast, "Nuclear Energy in Industry: Application to Oil Production," Climate Change and Energy Options Symposium, 1999, http://www.cns-snc.ca/events/CCEO/nuclearenergyindustry.pdf.
15. D. Ebner and S. Tuck, "Oil Sands players eye nuclear option: Soaring price of natural gas has producers looking for an alternative," *Globe and Mail*, September 23, 2005, http://www.theglobeandmail.com/servlet/ArticleNews/TPStory/LAC/20050923/ROILSANDS23/TPBusiness/?query=nuclear.

16. D. Ebner, "Husky Mulls Nuclear Option in Oil Sands," *Globe and Mail*, January 8, 2007, http://www.theglobeandmail.com%2Fservlet%2Fstory%2FRTGAM.20070108.whuskyy010 8%2FBNStory%2FBusiness%2F&ord=276943&brand=theglobeandmail&force_login=true.

17. Energy Alberta Corporation, 2007, http://www.energyab.com/.

18. *World Nuclear News*, "Not 'if' but 'when' for nuclear oil sands," 2007, http://www.world-nuclear-news.org/newNuclear/090107Not_if_but_when_for_nuclear_oil_sands.shtml.

19. CBC North News, "Company Eyes Slave rapids for hydro dam," February 27, 2006, http://www.cbc.ca/canada/north/story/2006/02/27/slave-dam-27022006.html.

20. Canadian Parks and Wilderness Society, Edmonton Chapter, "A Response to the Mineable Oil Sands Strategy," 2005, http://www.cpaws-edmonton.org/factsheets/CPAWS_Response_to_MOSS.pdf.

21. D. Struck, "Canada Pays Environmentally for U.S. Oil Thirst," *Washington Post*, May 30, 2006, http://www.washingtonpost.com/wp-dyn/content/article/2006/05/30/AR2006053001429.html.

22. J. Simpson, "Call a halt, Albertans," *Toronto Star*, July 7, 2006, http://thestar.workopolis.com/servlet/Content/fasttrack/20060707/COSIMP07?section=Energy; P. Lougheed, "Q&A with Peter Lougheed: Sounding an Alarm for Alberta," *Policy Options*, September 2006, http://www.irpp.org/po/archive/sep06/lougheed.pdf.

23. R. Schneider and S. Dyer, *Death by a Thousand Cuts: Impacts of In Situ Development on Alberta's Boreal Forest* (Calgary: Pembina Institute for Appropriate Development, 2006), http://pubs.pembina.org/reports/1000-cuts.pdf.

24. Human Resources and Skills Development Canada et al., Memorandum of Understanding for the Entry of Temporary Foreign Workers for Projects in the Alberta Oil Sands, 2004, http://www.hrsdc.gc.ca/en/epb/lmd/fw/mouforOilAlberta.pdf.

25. Suncor Energy Ltd., *12th Annual Progress Report on Climate Change—2006* (Calgary: Suncor Energy, 2006), http://www.suncor.com/data/1/rec_docs/571_Climate_Change.PDF.

26. Pembina Institute, *Annual Report, 2007*, http://www.pembina.org/about/annual-report.

27. D. Bueckart, "Environmentalists ask opposition to defeat government if Kyoto abandoned," *Montreal Gazette*, 2006, http://www.canada.com/montrealgazette/story.html?id=78a87d29-acbb-4bc2-8578-7dc5f66eb95e&k=68483.

28. Pembina Institute. "Wind Power by Pembina, 2007, http://www.pembina.org/wind/wind_power.php.

29. Pembina Institute, *Counting Canada's Natural Capital: Assessing the Real Value of Canada's Boreal Ecosystems* (Calgary: Pembina Institute for Appropriate Development, 2005), http://pubs.pembina.org/reports/Boreal_FINAL.pdf.

30. Pembina Institute, *Oil Sands Fever* [video], 2005, http://www.oilsandswatch.org/videos/osf.

31. C. Goodale et al., "Forest Carbon Sinks in the Northern Hemisphere," *Ecological Applications* 12, 3 (2002): 891–99; W. Kurz and M. Apps, "A 70-Year Retrospective Analysis of Carbon Fluxes in the Canadian Forest Sector," *Ecological Applications* 9, 2 (1999): 526–47.

32. W. Kurz and M. Apps, "An Analysis of Future Carbon Budgets of Canadian Boreal Forests," *Water, Air, and Soil Pollution* 82 (1995): 321–31.

33. M. Anielski and S. Wilson, *The Real Wealth of the Mackenzie Region: Assessing the Natural Capital Values of a Northern Boreal Ecosystem* (Ottawa: Canadian Boreal Initiative, 2007), http://borealcanada.ca/pdf/Embargoed_CBI_Mackenzie_Report_ENG.pdf.

34. D. Ebner, "Boreal Forest Said Worth More Than Diamonds," *Globe and Mail*, 2007, http://www.anielski.com/Documents/Boreal%20more%20than%20diamonds.pdf.

35. Canadian Boreal Initiative, 2007, http://www.borealcanada.ca/index_e.cfm.

36. Report on Business TV, "SqueezePlay," 2005, http://www.borealcanada.ca/whatsnew_docs/ROBTV_Squeezeplay_Dec%2013.mpg.

37. H. Brubaker, "Emerging source of oil, not from wells, but sand," *Philadelphia Inquirer*, 2004, http://www.energybulletin.net/3458.html.

38. E. Butterworth, "Partners for the Forest," Ducks Unlimited Magazine, 2004, http://www.ducks.org/DU_Magazine/DUMagazineNovDec2004/1229/Conservation.htm.

39. Pew Charitable Trusts, "The Pew Charitable Trusts: Serving the public interest," 2007, http://www.pewtrusts.com/.

40. World Wildlife Fund, "World Wildlife Fund Canada Welcomes Government Support of 'Conservation First' in the Mackenzie Valley," 2004, http://www.wwf.ca/NewsAndFacts/NewsRoom/default.asp?section=archive&page=display&ID=1366&lang=EN.

41. Parks Canada Agency, "Canada's New Government Takes Significant Step to Protect Major National Historic Site in the North," 2007, http://www.pc.gc.ca/apps/cp-nr/release_e.asp?id=1078&andor1=nr.

42. Canadian Parks and Wilderness Society et al., *Managing Oil Sands Development for the Long Term: A Declaration by Canada's Environmental Community*, 2005, http://pubs.pembina.org/reports/OS_declar_Full.pdf.

43. J. St. Clair, *Been Brown so Long, It Looked Like Green to Me: The Politics of Nature: Books* (Monroe, ME: Common Courage Press, 2004), http://www.amazon.com/Been-Brown-Long-Looked-Green/dp/1567512585; M. Dowie, *American Foundations: An Investigative History* (Cambridge, MA: MIT Press, 2002); F. Pace, *Wilderness, Politics, and the Oligarchy: How the Pew Charitable Trusts is Smothering the Environmental Movement* (Counterpunch, 2004), http://www.counterpunch.org/pace10092004.html; A. Cockburn and K. Silverstein, *Washington Babylon* (New York: Verso, 1996); M. Dowie, *Losing Ground: American Environmentalism at the Close of the Twentieth Century* (Cambridge, MA: MIT Press, 1996).

44. R. Sanders, "Howard Pew—Facing the Corporate Roots of American Fascism," *Press for Conversion!* 53 (2004), http://coat.ncf.ca/our_magazine/links/53/pew.html.

45. Canadian Boreal Initiative, "Boreal Leadership Council," 2003, http://www.borealcanada.ca/framework_blc_e.cfm.

46. D. Broten, "The Great Boreal Debate: Dances with Corporados," *Watershed Sentinel* 14, 1 (2004): 3–4, http://www.watershedsentinel.ca/Archives_WSS/ws141.pdf.

47. E. May and C. Pope, "An Open Letter to the CEO of Exxon Mobil," November 2005, http://www.sierraclub.ca/national/programs/atmosphere-energy/energy-onslaught/exxonmobil-letter-nov2005.pdf.

48. Sierra Club of Canada, "Moratorium on New Tar Sands Development Sought," 2006, http://www.sierraclub.ca/national/media/item.shtml?x=933.

49. Sierra Club of Canada, "Environmental Groups call on Imperial Oil to invest in a sustainable energy future," 2007, http://www.sierraclub.ca/national/media/item.shtml?x=1105.

50. D. Woynillowicz, *Tar Sands and Forests*, 2006.

51. H. McCullum, *Fuelling Fortress America: A Report on the Athabasca Tar Sands and U.S. Demands for Canada's Energy* (Ottawa: Canadian Centre for Policy Alternatives, 2006), http://www.polarisinstitute.org/files/Fuelling_Fortress_America-5_0.pdf.

52. Hewlett Foundation, "The Environment Program," 2004, http://www.hewlett.org/NR/rdonlyres/FF1455D1-708D-44D3-8071-9AAB693EA785/0/Environment04.pdf.

53. National Commission on Energy Policy, "Ending the Energy Stalemate: A Bipartisan Strategy to Meet America's Energy Challenges," 2004, http://www.energycommission.org/files/contentFiles/report_noninteractive_44566feaabc5d.pdf.

54. National Energy Policy Group, *National Energy Policy* (Washington, DC: U.S. Government Printing Office, 2001), http://www.whitehouse.gov/energy/National-Energy-Policy.pdf.

55. E. Kinetz, "A world without easy oil: What now?" *International Herald Tribune*, December 2, 2005, http://www.iht.com/articles/2005/12/02/yourmoney/mround03.php.

56. I. Mintzer, J. Leonard and P. Schwartz, *U.S. Energy Scenarios for the 21st Century* (Washington, DC: Pew Center for Global Climate Change, 2003), http://www.pewclimate. org/docUploads/EnergyScenarios.pdf.
57. Set America Free Coalition, "Set America Free," 2007, http://www.setamericafree.org/.
58. 109th Congress of the United States of America, *Energy Policy Act of 2005*, http://frwebgate. access.gpo.gov/cgi-bin/getdoc.cgi?dbname=109_cong_bills&docid=f:h6enr.txt.pdf.
59. Oil Sands Experts Group, Workshop: "Security and Prosperity Partnership of North America," Houston, Texas, January 24–25, 2006, http://www.fossil.energy.gov/programs/oilgas/publica- tions/oilgas_generalpubs/oilsands_spp_report_2.pdf.
60. F. Stonor Saunders, *The Cultural Cold War: The CIA and the World of Arts and Letters* (New York: New Press, 2001), http://www.amazon.com/gp/product/1565846648/102-1441743- 9684926?v=glance&n=283155.

Power Speaks to Power under the Nuclear Revival Tent

by Marita Moll

On January 17, 2007, the Doomsday Clock was moved from seven minutes to five minutes before midnight—midnight symbolizing the figurative end of civilization. Created in 1947, the clock has been moved seventeen times in response to the world's perceived vulnerability to nuclear-based threats. In 1953, it was set at two minutes to midnight when the US and Russia, deep into an arms race, began testing hydrogen bombs in the Pacific Ocean. In 1991, with the Cold War officially over and various treaties in place to control nuclear proliferation, the clock was set at seventeen minutes to midnight. The world was optimistic that the nuclear genie could be, if not put back into the bottle, at least put on a very tight leash.

Currently, there are 27,000 nuclear weapons around the world, 2000 of them ready to launch within minutes, and nuclear aspirations in political hotspots like Iran and North Korea. It is reason enough for global warnings. But this time there is a new twist. With Japan planning to build five new nuclear plants by 2010 and China planning thirty reactors by 2020, civilian nuclear power is ramping up to levels never dreamed of before. In addition, existing reactors around the world are aging and energy demands are rising. It all adds up to substantial risk for current and future generations. As reported by the group responsible for the clock, the board of directors of the *Bulletin of the Atomic Scientists*,

> While nuclear energy production does not produce carbon dioxide, it does raise other significant concerns, such as the health and environmental hazards of nuclear waste, the production of nuclear materials that can be diverted to the production of weapons, and the safety and security of the plants themselves.[1]

This prestigious scientific community, which includes 18 Nobel laureates and as well as many other notable scientists, including Stephen Hawking, is clearly not buying the argument that civilian nuclear power is a way out of the climate-change crisis.

On January 17, 2007, Canadian natural resources minister Gary Lunn was promoting a radically different point of view. "Nuclear energy is emission-free, there's no greenhouse gases, there's no pollutants going out [with] the energy," he told a news conference held to announce the new federal ecoEnergy Technology Initiative. "There's a great opportunity to pursue nuclear energy, something I am very keen on," said Lunn.[2] The contrast could hardly have been more pronounced: on the one side, the world's most distinguished scientists carefully weighing the evidence and warning against nuclear expansion; on the other side, "Canada's New Government" strutting its stuff on its first "keen to be green" coming-out party. Unfortunately, promoting and supporting this problematic but powerful industry is still a very big part of Canadian public policy.

Ontario "Nukes of Hazard"

The nuclear establishment in Canada works hand in hand with legislative and regulatory bodies at all levels. "The nuclear power industry—whose boards and senior personnel are staffed with well-paid political cronies—is coming back, not because it has become safe or economical or reliable, but because of the close relationship nuclear directors and CEOs have with the people in power," says Tom Adams, executive director of Energy Probe, an energy watchdog group, who is very familiar with this insider network.[3] The Ontario government's 2006 decision to submit the refurbishment of four of Ontario's oldest reactors to the lowest level of environmental assessment available under federal law exemplifies this pattern. It has the politics of nuclear power written all over it.

Ontario was an early adopter of nuclear technology with the first power stations coming online in 1971. There are currently three nuclear power generating stations in Ontario (Bruce, Darlington and Pickering) with sixteen functioning reactors between them supplying 50 percent of the province's electrical-energy needs. Sitting right on Toronto's doorstep is Pickering, the oldest of these power stations, the largest nuclear power station under one roof in the world and unique in its proximity to a major population centre. Pickering A and Pickering B contain four CANDU pressurized heavy water reactors each.

The CANDU reactors are supposed to last forty years; but at about the twenty-year mark, they start running into difficulties. This is particularly true with the feeder pipes that circulate coolant and the pressure tubes that hold uranium bundles. These pipes and tubes have deteriorated faster than anticipated, and replacing them is proving to be a very costly undertaking. In 1997, all four Pickering A reactors had to be taken offline because of problems in this area. Two reactors were eventually restarted at five times the estimated cost. Although there were two review committees struck during this period to inquire into the financial and technical problems at Pickering, the appointed committees consisted of ex-politicians like Jake Epp, well-known pro-nuclear energy minister under Brian Mulroney and others unlikely to rock the nuclear boat. A better energy plan that would have ramped up sustainable alternatives and conservation strategies back in 1997 never emerged. Instead, $4 billion was spent to rebuild two old reactors. Interestingly, John Baird, the Ontario environment minister at the time, is now the federal environment minister. As Adams notes, the relationships between environmental and resource industry politicians and the nuclear industry are very close indeed.

Greenpeace Canada sums up the unintended consequences of this badly managed plan:

> To keep the lights on, Ontario was forced to boost production at its coal-fired generation stations, increasing greenhouse gas emissions by 120%... The long delays and eventual failure of the Pickering A restart resulted in significant climate change and smog causing emissions—a significant environmental impact.[4]

Nuclear is clearly a poor ally in the war against climate change.

In June 2006, the Ontario Liberal government, under pressure to phase out coal-fired power stations, announced a new twenty-year power plan which once again leaned heavily on nuclear power generation. Having made a decision to refurbish the four Pickering B reactors, the government's first move was quietly to pass a regulation that exempted the project from an assessment under the Ontario *Environmental Assessment Act*. In its place was substituted an environmental screening process by the Canadian Nuclear Safety Commission (CNSC)—a process that does not allow interveners to question government policy or cross-examine experts or consider "alternatives," but rather restricts itself to assessing the technical aspects of the specific project under review.

The few interveners attending the January 24, 2007, CNSC public hearing on the proposed guidelines for the environmental impact study were unanimous in their demand: they wanted a full panel review under the *Canadian Environmental Assessment Act*—a review that would look at all the implications of the proposed ten-year, multi-billion-dollar undertaking. After all, it was pointed out, under today's rules, a nuclear station in the Pickering location would be out of the question. But most concerns raised by interveners were either considered outside the scope of the hearings or "manageable risks" by the proponent, Ontario Power Generation (OPG), and the CNSC staff who saw remarkably eye-to-eye on most issues. Despite assurances from the chair that there were no foregone conclusions, interveners were highly skeptical that the hearing would lead to any broadening of guidelines already agreed to between OPG and CNSC staff.

In the final analysis, nuclear power in Ontario is still a monster with its hands firmly on the rudder of power planning and implementation, despite the provincial government's public commitment to support renewable energy sources. With various formal structures and processes in the system actively supporting the industry, the refurbishment of Pickering B and any other nuclear expansion in Ontario has very little to do with climate change or environmental responsibility and everything to do with the politics of nuclear power.

Nuclear Power Won't Clean Up the Oil Sands

The very biggest environmental issue in Canada is the Alberta oil-sands project. Though as yet only partially developed, it covers 138,000 square kilometers of northern Alberta—an area as large as the state of Florida. At the moment, only the surface oil is being extracted—about 3 percent of the total. But already, the environmental degradation has been colossal. Giant machines dig up two tonnes of oil sands for every barrel of oil produced. This area is a bleak moonscape of open pit mines and giant lakes of toxic sludge visible from outer space. The oil or bitumin is separated from the sand by steaming it out with water heated by gas-fired generating stations (2 to 4.5 cubic metres of water are needed to extract 1 cubic metre of synthetic crude oil). Getting at the deepest pockets of oil will be even more energy intensive.

As a direct result of this project, Alberta has some of the highest per capita greenhouse gas (GHG) emissions in the world, three times higher than the Canadian average—and six times greater than western Europe. Between 1990 and 2002 alone, Alberta's GHG emissions increased by 29 percent.[5] Since the National Energy Policy uproar in the 1970s, few federal government officials have dared challenge the Alberta oil industry. But recently, even the western-based Harper government has been suggesting that GHG emissions cannot continue to escalate at this rate.

Having waited a long time, this is precisely the moment at which the nuclear industry wants to step up to the plate. In October 2006, Atomic Energy of Canada Ltd. (AECL) announced that it had signed a two-year exclusive deal with Calgary-based Energy Alberta Corporation to establish the CANDU reactor technology in the oil sands.[6] Energy Alberta Corporation says it has a number of companies willing to buy power from them. They are currently looking for appropriate building sites and expect that reactors would be ready for operation in eight years.

Gary Lunn, Canada's natural resources minister, is actively promoting the deal. "It's not a question of if, it's a question of when in my mind," said Lunn in an interview with Sun Media. "I think nuclear can play a very sig-

Local News:

Bringing you the day's top stories and events from around the province and the country.

Weather:

Find the current weather conditions where you live.

Community Calendar:

Check out what's happening in your community and submit your events online.

Programs and Special Features:

Learn more about your favorite radio and television programs and hosts, and read our in-depth features.

cbc.ca/manitoba

cbc.ca ⊛

nificant role in the oil sands. I'm very, very keen," he continued, admitting that he had already been involved in discussions on the issue.[7] The involvement of AECL, a federal crown corporation which has received $17 billion in subsidies over the last twenty years to build and promote CANDU reactor technology, gives the federal government a solid stake in the outcome of this discussion.

In January 2006, there was a two-day meeting in Houston, Texas, between US and Canadian oil executives and government officials. The meeting was organized by Natural Resources Canada and the US Department of Energy. They were discussing a five-fold expansion in oil-sands production in a relatively short time span and were also discussing streamlining environmental assessments. Although the federal government has denied it would ever consider this, with AECL now in the game, that denial may be shortlived. Alberta environmental groups are gearing up for a fight. In December 2006, the Pembina Institute released a 130-page report *Nuclear Power in Canada: An Examination of Risks, Impacts and Sustainability*.

Meanwhile, there is a nuclear revival underway in neighbouring Saskatchewan. Saskatchewan is the major supplier of uranium to nuclear sites around the world. With the oil-sands project nearly on its doorstep and coming closer, various stakeholders are looking for a share in the bonanza. In January 2006, the University of Regina along with Saskatchewan rural and urban municipalities associations held an industry-dominated conference called "Exploring Saskatchewan's Nuclear Future." Among the guest speakers was Patrick Green, former Greenpeace activist, now promoter of nuclear energy as a response to climate change.

Clearly, there is considerable momentum on the side of proponents of nuclear expansion into the oil sands. Supporting an already environmentally destructive project by bringing in another environmentally destructive project would exacerbate the problem. Not only would the land be further raped and pillaged, but future generations would have to deal with thousands of years of nuclear waste as well. And, in the long run, it will buy nothing but a few extra years of oil dependency before we will have to get serious about exploiting alternative energy sources anyway.

Stocking Up On Nuclear Waste

Amazingly, even though fifty years of study has not yet solved the waste-storage issue, Canada continues to support the nuclear business. We have already stockpiled 2 million spent fuel bundles (about 45,000 metric tons)—most of it in Ontario—an amount that would fill up five hockey rinks from the ice surface to the top of the boards. Canada's current nuclear reactors will produce 3.6 million bundles during the course of their average operation life of forty years, which will fill another five hockey rinks. This spent fuel contains over one hundred different radioactive isotopes. While it is true that some of these dissipate quickly, others, like plutonium-239 (weapons-grade plutonium), which has a half-life of 24,000 years, will still be around to poison the environment many thousands of years from now.

The Nuclear Waste Management Organization (NWMO) was set up in 2002 to study this problem. Its shareholders are the nuclear industry itself: Ontario Power Generation, Hydro Quebec and New Brunswick Power. Critics point out that a group with a vested interest in continuing the production of such waste should not be in charge of deciding what to do with it. It is unlikely, however, that they will be inclined to look for ways to phase themselves out.

In its final report, the NWMO recommended that Canadian waste be stored in a deep underground facility, most likely in the Canadian Shield. This waste would be monitored and remain retrievable over time, with the estimated cost of disposing of it being $24 billion.

But, as already noted in the discussion on the refurbishment of Pickering reactors, the limited scope of these processes and the incestuous power relationships involved determine their outcomes. "The NWMO must examine the whole cycle of nuclear waste, from production to disposal," says John Bennett, Sierra Club senior policy advisor on atmosphere and energy. "To not do so is looking at the problem without attacking the cause, and allows the nuclear industry to continue to push for the construction of uneconomic, unsustainable nuclear power plants."[8]

According to a January 2007 CBC radio report, natural resources minister Gary Lunn has been reviewing the NWMO plan and will soon take it to cabinet with a decision possible within six months. There are many challenges ahead, however, notes Pembina Institute's Mark Winfield. These include the very explosive transportation issue. For the communities involved, "it's estimated that there would be two to three truckloads a day, every working day for 30 years, to move material to a central storage facility."[9] Luckily, Canadian communities are rarely complacent on this issue. This is a political hot potato and guaranteed to be a long and drawn out battle that a responsible government would do best to avoid.

Nuclear in a Post 9-11 World

The culture of secrecy around nuclear technology enables the current revival by avoiding the big issues. While major accidents are hard to conceal, smaller accidents, what the industry prefers to call "incidents," happen regularly. In 1983 there was a very serious accident at Pickering involving a pressure tube that burst without warning in the core of the reactor. In 1997, an outside review of Ontario's reactors found that the safety margins were "minimally acceptable"—the lowest rating before mandatory shutdown.[10] Insurance companies around the world refuse to insure property against nuclear accidents. Should there be an accident, nuclear power plant owners' liability is limited to about $415 million in today's dollars. In July 2007, The Federal Government Increased That Amount To $615 million—about 10 cents per dollar of dwelling value around the Pickering plant according to Tom Adams of Energy Probe.[11] The Canadian Nuclear Society itself admits that its relationship with the public is "fragile." It takes only one accident or potential accident to freeze out the nuclear industry for a very long time.

In an age of increasing fear of terrorist attacks, security has become an acute concern. According to a *Toronto Star* report, in January 2007, the CNSC informed the Canadian nuclear industry that all new reactor designs must meet international safety standards with respect to their ability to withstand a massive outside shock or explosion—such as the impact of a commercial jetliner.[12] Many of the interveners at the CNSC hearing on the proposed Pickering refurbishment were asking, "why the double standard?" Pickering and other aging nuclear stations are clearly vulnerable sites. Citing security issues, the CNSC will not engage in this discussion in hearings, but there is a visible nervousness every time the subject comes up.

Is nuclear power a legitimate response to the crisis of climate change? Too many indicators suggest that it is not. In its thorough nuclear life-cycle analysis, the Pembina Institute has stated:

> While GHG emissions associated with nuclear power are less than those that would be associated with conventional fossil fuel energy use, no other energy source combines the generation of a range of conventional pollutants and waste streams with the generation of extremely large volumes of radioactive wastes that will require care and management over hundreds of thousands of years.[13]

For many years, the world ignored warnings about how our human activities are affecting the earth's atmosphere. It is already too late to undo some of this damage. Do we now intend to ignore warnings about the dangers of the unbridled expansion of nuclear power plants to feed our ever-increasing demand for electricity? Radioactive waste, safety and security risks, as well as the unacceptably high amount of GHGs produced all amount to too high a price to pay for what we like to call progress. The only genuine response to climate change is to put all possible money and talent into pursuing sustainable alternative energy sources. In this scenario, nuclear energy has no place.

Notes

1. *Bulletin of the Atomic Scientists*, "Doomsday Clock Moves Two Minutes Closer to Midnight," http://www.thebulletin.org/weekly-highlight/20070117.html, accessed January 18, 2007.

2. Canadian Broadcasting Corporation, "Tories to Spend $230M on Clean Energy Technology," http://www.cbc.ca/technology/story/2007/01/17/clean-energy.html, accessed January 17, 2007.

3. Tom Adams, "Expose the Truth about the Nuclear Industry," http://www.energyprobe.org/energyprobe/index.cfm?DSP=content&ContentID=10223, accessed April 28, 2007.

4. Canadian Nuclear Safety Commission, Submission from Greenpeace Canada in the Matter of Ontario Power Generation Inc. Oral Presentation, January 24, 2007.

5. Nicola Ross, "Time for an Oil Change: Alberta Should Tap its Much-Lauded Entrepreneurial Spirit to Lead the Way in Emission Reduction," *Alternatives Journal,* http://www.accessmylibrary.com/coms2/summary_0286-13986418_ITM, accessed January 1, 2007.

6. Atomic Energy of Canada Limited (AECL), "CANDU Viable Option for Energy Dependent Oil Sands Production," http://www.aecl.ca/NewsRoom/Bulletins/E-061002/E05.htm, accessed October 2, 2006.

7. Alan Findlay, "Nukes Aimed at Oil Sands," http://cnews.canoe.ca/CNEWS/Canada/2006/12/21/2914252-sun.html, accessed December 21, 2006.

8. Sierra Club of Canada, "NWMO Report Result of Mandate Twisting," news release, http://www.sierraclub.ca, accessed November 4, 2006.

9. Canadian Broadcasting Corporation, "Nuclear Waste Disposal Plan Could Come within Months," http://www.cbc.ca/canada/story/2007/01/17/nuclear-waste.html, accessed January 17, 2007.

10. Elaine Dewar, "Nuclear Resurrection," *Canadian Geographic*, May–June 2005, p. 74. 11. Tom Adams, "The nuclear shield," *Globe and Mail*, July 16, 2007.

12. Tyler Hamilton, "Could Reactors Withstand Blast?" *Toronto Star,* January 19, 2007.

13. Mark Winfield et al., *Nuclear Power in Canada: An Examination of Risks, Impacts and Sustainability* (Pembina Institute, 2006).

Capturing Revenues from Resource Extraction

by John W. Warnock

Natural resources are a free gift from nature. In all human societies for thousands of years natural resources were considered to be common property available to all for equal use. As humans moved to horticultural societies, families were often granted use rights to certain resources like land for farming or grazing animals or special fishing sites. But these rights were never considered private property and could be withdrawn by society.

The new capitalist system required the privatization of natural resources so that they could be bought and sold in a market. The European imperial powers imposed this new system of resource ownership on the areas of the world that they colonized. This was one of the fundamental aspects of European imperialism and colonialism in the nineteenth century: the world-wide transformation of common land and natural resources into private property.

Economic Rent under Capitalism.

The concept of rent that we use today has its roots in the ideological defence of private property in resources constructed by the earliest political economists. John Locke (1637–1704) first set forth the classic case that was the foundation for all others, defending England's seizure of land from the indigenous peoples of the Americas. The early capitalists insisted that any individual or business could seize land and resources that were not being efficiently used to produce a profit. In doing so they owed nothing to the population in general for this action.

The most widely cited liberal definition of economic rent was set forth by David Ricardo (1772–1823). This is the concept presently used by economists, resource industries and governments. Economic rent is the surplus that is created by the use of a natural resource over and above what is necessary to keep labour and capital on the land and producing products. These costs include a normal profit. Under a condition of perfect competition, there is no economic rent. Economic rent is created only when the exploitation of a natural resource like oil or gas produces a return that is over and above the normal rate of return in a competitive market. It is a *monopoly profit* or an excess profit.

Under this liberal capitalist view of economic rent, Ricardo did not include the payment of compensation or a royalty to the general public for the privatization and use of natural resources that had previously been considered social or public property. But with the expansion of political democracy, governments began to demand a compensation for the use of natural resources. These are commonly called royalties and are considered a basic cost of production by all resource extraction industries.[1]

The Oil and Gas Industry

Within the oil and gas industry today, economic rent is generally defined as the difference between the cost of exploration, field development, extraction, royalties and fees and the market price. These costs include a normal rate of return on investment. While the normal return on equity in Canada is around 4.5 percent, the oil and gas industry insists that a 12 percent return is needed to attract capital investment because of various "risks." In this industry there has always been a very large economic rent and an ongoing political struggle over what share of that surplus profit should go to the private corporations and what share should go to the government.

Rent is also difficult to measure because the oil industry has never operated in a free market. It has always been characterized by monopoly, oligopoly and government support. The large private integrated corporations, the Seven Sisters, operated as a formal cartel for decades. The super majors today do not compete with one another but work together to support their common interests. In most areas of the world they operate joint development ventures with each other.

The oil and gas corporations have been supported by governments that accept oligopoly and provide a wide range of economic assistance. For years the oil industry in the United States was supported by import quotas and a pro-rationing system that controlled production, both designed to prevent the free market from working and to maintain stable prices.

Since the creation of OPEC the super majors and their supporters in government have worked closely with Middle East countries to develop a market system where oil sells within a prescribed band of prices. This guarantees a good revenue to both the producing governments and the corporations. As a practical matter, the basic cost of production has been set to maintain a viable industry in the United States. This has resulted in very high economic rent in low cost production areas, particularly the Middle East and North Africa.[2]

The key to this is the ongoing relationship between the Royal Family in Saudi Arabia and the US government. To illustrate this fact, in 2005 the average conventional oil well in the United States produced eleven barrels of oil per day, in Canada eighteen barrels and in Saudi Arabia 5800 barrels.[3]

Fiscal Systems for Extracting Economic Rent.

All governments wish to obtain at least some revenues from the extraction and use of renewable and non-renewable resources. These resources are commonly owned by the state, and governments have the responsibility for controlling their extraction and use. For non-renewable resources like oil and gas, they are depleted and once gone cannot be replaced. Thus it has been argued that governments have a moral responsibility to ensure that future generations benefit in some way from their depletion. For countries moving out of an agriculture economy (including Canada), natural resources or staple industries have been a way of diversifying the economy, creating jobs and developing backward and forward links to industries that provide inputs and process the raw materials. In many cases, including Saskatchewan, these resource extraction and industrial projects can bring economic development and employment to rural and remote communities.

Therefore, in a democratic society, governments should be seeking to maximize their share of the economic rent, or excess profits, over the life of any resource development project. All governments also want to have steady and predictable revenue so that they can plan for government expenditures. A democratic government would be expected to place a high priority on developing policies that guarantee that the economic rent is re-invested in the local area or province. If this does not happen, resource development results in boom and bust communities. A socially responsible democratic government also aims to have a large share of the benefits from resource development accruing to local indigenous populations.

Royalties and Fees in the Oil and Gas Industry

Within the Western capitalist oil and gas industry it is assumed that private corporations and governments each have a right to a share of the economic rent (or surplus profit). Economic rent from oil and gas developments will vary depending on the size of the resource deposit, its grade, its ease of extraction, its location, the state of the local infrastructure and the distance from important markets.

Corporations wanting to develop oil or gas in Canada must pay fees that enable them to gain access to land for prospecting, exploration and leases for production. Production royalties are imposed and take the form of a fixed percentage of the gross revenue from extraction. In the early concession agreements in the Middle East, the royalty paid to the government by the independent oil companies (IOCs) was set at 12.5 percent of the gross revenues. In Canada, *ad valorem* royalties have different scales according to the time when the resource came into production. Old oil, which was less costly to develop, has a higher royalty than new oil, which is found in smaller pools and at deeper levels.

In order to stimulate production, governments have on occasion introduced very low royalties. The Venezuelan government in the late 1990s lowered the royalty rate to only 1 percent. The government of Alberta has set the royalty rate in the tar stands at only 1 percent until the corporations have recovered all of their capital investments.[4]

In the 1970s Allan Blakeney's NDP government in Saskatchewan raised royalties significantly in order to capture a greater share of resource revenues from the extraction of oil and gas. Between 1976 and 1982 royalties averaged 52 percent of oil sales. This has been reduced to only 16 percent under the present NDP government headed by Lorne Calvert. The current figure is not out of line with royalties charged around the world. But it is widely understood that royalties are not the appropriate government tool to capture economic rent.[5]

Capturing Economic Rent: An Example from Saskatchewan

In Saskatchewan reserves of light crude and medium crude are in steady decline, as is the case across the Western Canada Sedimentary Basin (WCSB). The decline in these valuable sources is being replaced by heavy crude, which is more costly to extract and process. Technological innovations, including horizontal drilling, have actually reduced the costs of drilling wells on the Canadian prairies. The industry also benefits from federal and provincial subsidies and supports, very low royalties and tax rates, a good drilling environment, local refineries and a pipeline system to the most important world market.[6]

What market price is necessary to produce a viable oil industry on the Canadian Prairies? In 2004 the Canadian Energy Research Institute (CERI) reported that for new tar sands projects a "West Texas Intermediate price of US$25 per barrel would enable an oil sands project developer to cover all costs and earn a 10 percent return on investment." A year earlier the National Energy Board concluded that "a price of US$22 per barrel provides adequate returns to support investment in the oil sands and offshore oil development."[7]

The cost of extracting heavy oil in Saskatchewan is below that incurred in the off-shore and tar-sands industries. In 2003 Li Ka-shing, the Hong Kong owner of Husky Oil, stated that the Lloydminister Heavy Oil Upgrader was profitable when the price of West Texas Intermediate (WTI) light crude was $18 per barrel. In 2002 Saskatchewan Energy and Mines reported that "reasonable levels of conventional activity can be maintained at the WTI price in excess of US$20 per barrel."[8]

In 2005 the average price of light and medium crude extracted in Saskatchewan was $68 per barrel. But heavy oil has a discount price from light crude oil, in recent years around 30 percent. The price of Hardisty Crude Oil at Bow River (heavy oil) has gone from $25 a barrel in 2001 to $38 in 2004 and $46 in 2005. So how much economic rent is captured from the extraction of heavy oil in Saskatchewan?

It is very difficult to determine the profitability of the oil and gas industry. Their data are not made public. They also have a range of methods of transfer pricing, plus all the major corporations use off shore tax havens. The public is forced to depend on the data provided by the Canadian Association of Petroleum Producers (CAPP) or the US Energy Information Administration. The cost figures provided by CAPP are significantly above those provided by the corporations to the US Department of Energy.

One recent study by ARC Financial Corporation using CAPP data reports that the average annual operating cost of conventional wells in the WCSB in 2006 was around $7 per barrel.[9] To this is added $2 per barrel for general and administrative expenses. In addition, royalties and bonus fees averaged $8 per barrel. (There are provincial and federal taxes on profits and income, but these apply to all corporations and cannot be seen as a method of collecting economic rent.)

Report on Business magazine records that for 2003 the big ten Canadian oil companies reported a return on equity that ranged between a low of 18 percent for Talisman Energy to a high of 34 percent by Imperial Oil. The very high return on equity reflects the fact that almost all of the economic rent produced in recent years in Canada has gone to the private sector.

Looking at the case of heavy oil in Saskatchewan, the report by ARC Financial Corporation suggests that extraction costs were around $9 per barrel, royalties and bonus fees were $8 per barrel and return on equity would be well above the cost of capital. This confirms the statements by Husky Oil and the Saskatchewan Department of Industry and Resources. Of the $46 average price for heavy oil in 2006, at least $26 per barrel would be economic rent, and all of this went to the private sector.

Rent Collection in the Present World Situation

Over the past few years we have seen the world price for oil triple, going as high as $78 per barrel in 2006. This has greatly increased the economic rent produced by this industry. While private industry seeks to maximize its share of these windfall profits, most governments in oil producing countries have moved to try to capture most of this for their own citizens.

In the Middle East, the OPEC countries have state owned oil corporations (known in the industry as National Oil Companies or NOCs). This allows them to capture all of this rent. Where they have joint ventures or production sharing agreements with private oil companies, they have had term contracts that specify that when the international oil price increases the share going to the government increases. This has allowed them to capture all of the increases in economic rent in the past three years. The United Arab Emirates has a "make-up tax" that limits the margin of profit of the private corporations to $1 per barrel. Full cycle production costs in Saudi Arabia have been estimated by the noted oil economist M.A. Adelman to be around $2.90 per barrel.[10]

Many oil producing countries employ production sharing agreements, where private oil companies must give a percentage of the oil extracted to the government. Kazakhstan requires all private companies to give 80 percent of the oil extracted to the government. In October 2005 Libya opened off-shore fields to private development by way of a closed bidding process. The seventeen companies who were successful agreed to give the Libyan National Oil Company between 70 and 94 percent of the oil extracted.[11]

Almost all of the oil producing countries have NOCs, and they engage in joint ventures with the private oil corporations. It is common for the NOC to hold 50 percent of the equity in the joint venture and thus receive 50 percent of the share of the oil revenues received. Under the government of Hugo Chavez in Venezuela, its NOC holds a minimum of 60 percent of the equity in all joint ventures.[12]

During this recent period of high windfall profits, a number of oil producing countries introduced some form of excess profits tax to try to capture these revenues. Russia now imposes an excise tax that takes 90 percent of the value of sales over $25 per barrel. In 2005–6, Ecuador imposed an excess profits tax on the oil industry, and when Occidental Petroleum refused to pay the back taxes, it was nationalized. Eight oil and gas corporations immediately volunteered to assume their production for the government. A new law grants the state 50 percent of all gross income above a fair market price, which is determined by the government.[13]

If a government is seriously concerned about gaining a fair share of the economic rent from resource extraction, then it must be directly involved in the industry. The examples of Enron in the United States and Yukos Oil in Russia demonstrate the ability of transnational corporations to use intra-corporate transfer pricing and off-shore tax havens to hide profits. The government of Alaska has on a number of occasions taken the oil companies to court for tax avoidance and won major settlements. They now use the universal taxation policy to undermine intra-corporate transfer policies.[14]

Governments that have NOCs obviously have the best chance of capturing economic rent. But it is not necessary to have a large, integrated state-owned oil corporation. Quite a few NOCs in smaller oil producing countries simply contract out the development work to the many service companies in the oil and gas industry. Before deregulation, SaskPower used this system to develop natural-gas reserves.[15]

Conclusion: The Failure of Canadian Governments to Capture Economic Rent

Over the past three years we have seen a dramatic increase in the price of oil which has been unrelated to the cost of production. Governments of producing countries have all moved to capture at least some of these windfall profits. But this has not been the case in Canada. Why is this?

In contrast to Canada, Norway has an extensive system for capturing economic rent, including a world-class state-owned company, Statoil, direct government equity in off-shore oil and gas developments, special petroleum taxes, carbon taxes and duties. The Norwegian government has managed to capture around 78 percent of the economic rent from the recent price increases.[16]

I would argue that Canada's governments have not chosen to follow the path of Norway because of the deep integration of the industry into the United States, the continental energy policy now entrenched in NAFTA, the presence of large transnational corporations that dominate the industry here and the fundamental government commitment to a neoliberal political economy since the early 1980s.

But there is more than just this. It seems to me that Canadian energy policy, accepted by all levels of government, is the result of the overall political commitment of Canada to the support of the Anglo-American political alliance to dominate the world. Canada is a member of NATO, NORAD and the OECD, and our governments support the Anglo-American position at the United Nations, the IMF, the World Bank and the WTO. Canada is also a special member of the Anglo-American political alliance (with New Zealand and Australia) for the development of nuclear weapons, the development and production of chemical and biological weapons and the secret intelligence agreement that now includes the advanced satellite system "Echelon." This requires that Canadian governments basically follow the political direction set by the governments of the United States and Great Britain.

Notes

1. John W. Warnock, *Saskatchewan: The Roots of Discontent and Protest* (Montreal: Black Rose Books, 2004).

2. F. William Engdahl, *A Century of War: Anglo-American Oil Politics and the New World Order* (London: Pluto Press, 2004); Michael T. Klare, *Resource Wars: The New Landscape of Global*

    Conflict (New York: Henry Holt, 2002); Oystein Noreng, *Crude Power: Politics and the Oil Market* (London: I.B. Tauris Publishers, 2002); Toby Shelley, *Oil: Politics, Poverty and the Planet* (London: Zed Books, 2005); Daniel Yergin, *The Prize: The Epic Quest for Oil, Money and Power* (New York: Touchstone Books, 1991).

3.   Michael C. Lynch, "Saudi Arabia's Oil Supply: Conjecture and Reality," *Petroleum Intelligence Weekly*, June 5, 2006, http://www.energyintel.com.

4.   Bernard Mommer, *Global Oil and the Nation State* (Oxford: Oxford University Press, 2002).

5.   John W. Warnock, *Natural Resources and Government Revenues: Recent Trends in Saskatchewan* (Regina: Canadian Centre for Policy Alternatives, Saskatchewan Division, 2005).

6.   Canadian Energy Research Institute, information session on conventional oil, Calgary, January 26, 2005, http://www.ceri.ca.

7.   Bob Dunbar, "Oil Sands Supply Outlook," CERI *Energy Insight* 2 (October 2004); Canada, National Energy Board, *Canada's Energy Future: Scenarios for Supply and Demand to 2025,* (Ottawa: Government of Canada, 2003).

8.   Saskatchewan Energy and Mines, *Oil in Saskatchewan*, 2002.

9.   Peter Tertzakian and Kara Baynton, *Canadian Upstream Oil and Gas Industry Financial Performance: Outlook 2006–2008* (Calgary: ARC Financial Corporation, 2006).

10.   Lynch, "Saudi Arabia's Oil Supply"; *Petroleum Intelligence Weekly*, various issues, 2005–6.

11.   *Petroleum Intelligence Weekly*, October 10 and 17, 2005.

12.   See www.venezueleananalysis.com, 2006.

13.   *Petroleum Intelligence Weekly*, various issues, 2005–6.

14.   Alaska Finance, http://www.tax.state.ak.us.

15.   John W. Warnock, *Selling the Family Silver: Oil and Gas Royalties, Corporate Profits, and the Disregarded Public* (Edmonton: Parkland Institute; Regina: Canadian Centre for Policy Alternatives, Saskatchewan Division, 2006).

16.   Ole Gunnar Austvik, "What Norway is Doing with Petroleum Rent Collection and Use," presentation to the conference, Power for the People, Parkland Institute, University of Alberta, November 18, 2006. Available at http://www.kaldor.no/energy.

The US Energy Act and Electricity
What it Means for Canada

by Marjorie Griffin Cohen

Canadians who care about the effects of US imperialism associate current threats with the uncompromising positions the US takes on trade deals and aggressive measures related to security. The drama of these events—particularly the war in Iraq, softwood lumber and beef trade disputes, border issues and US surveillance of Canadians—overshadows the less visible but equally insidious acts of US imperialism. I refer here to the regulatory changes the US makes by fiat on issues like energy that affect Canadian economic security. These involve actions that lead to a deep and permanent integration, with the US firmly in control.

Canadians did not pay much attention to the US *Energy Policy Act* (EPA) passed in the summer of 2005—aside, that is, from the warnings of environmentalists that the Act would likely enhance North American environmental destruction. Likely, the Canadian media and politicians ignored it because the American point of view already firmly dominates the oil and gas sectors in Canada. And the federal government is squeamish about anything that looks like it might be contemplating a national energy policy. But by adopting a head-in-the-sand approach to energy, Canada risks energy security for the future, particularly for one area of energy that is still mostly in public hands in Canada, namely electricity. The ways electricity in Canada is generated and sold is increasingly and speedily coming under the control of US initiated policy.

This chapter is about how changes in the electricity market and public policy in the US are affecting Canada's electricity future. The focus is on the US Federal Energy Regulatory Commission (FERC) and its attempts to redesign and control the North American electricity market. This involves a discussion of the effects of FERC's insistence on "reciprocity" in market access (that is, its requirement that whatever changes occur through their regulatory rulings must be reciprocated in any trading partners) and an appraisal of the new powers given to FERC through the new US Energy Bill.

US Energy Policy

US official policy is designed to promote private electricity through a competitive "North American" electricity market run by American rules and

American players. This was firmly articulated through the report of the National Energy Policy Development Group, headed by Vice-President Dick Chaney and US Secretary of State, Colin Powell (Cheney et al. 2001). This report recognized the significance of Canada for US energy supply and called for increased regulatory integration in electricity: "The reliability of the North American electricity grid can be enhanced yet further through closer coordination and compatible regulatory and jurisdictional approaches."[1]

"Compatible regulatory and jurisdictional approaches" may sound relatively benign and even reasonable to those who do not understand the fundamental differences between the Canadian and US systems of electricity production and distribution. But accepting US regulatory dominance in electricity will result in the Canadian system switching from one based on public utilities that engage in long-term planning to a competitive market-based model that will rely on the vagaries of the market and privatization of electricity to determine how much is produced and who gets it at what price.

When the Cheney/Powell report talked about "further" cooperation in regulatory issues it was recognizing that a great deal had already occurred that put Canada on the same course as the US with regard to energy market deregulation. Deregulation is the name for changing the system from one dominated by large utilities that were vertically integrated[2] and responsible for electricity in specific geographic regions to one that was designed with power trading in mind. The model for deregulation began with the break-up of the Bell telephone system in 1984 by a Supreme Court decision and continued rapidly through other industries.[3] For a variety of reasons, deregulation of electricity came relatively late and was largely the result of enormous pressure from private industry to have access to public transmission lines.[4]

Ensuring private access to utility transmission lines was the major focus of FERC activities to achieve deregulation at the outset. More recently, since George W. Bush came to power, FERC has become more imperialistic. It is now intent on establishing Standard Market Design (SMD) throughout North America. The *Energy Policy Act* of 1992 liberalized wholesale generation by exempting a whole class of generating facilities from regulatory control and, at the same time, granting them access to transmission systems. This marked the beginnings of constructing a system of long-distance trading in electricity.[5] These changes were further strengthened by FERC orders 888 and 889 in 1996, which removed monopoly powers over transmission from the utilities. In particular, it eliminated the sequential marking up of transmission charges that inhibited long-distance transactions.[6] Since utilities usually own the transmission systems, removing utility control over transmission is crucial to privatization initiatives. To achieve this in 2000, FERC issued order 2000 calling for the creation of very large transmission areas, or Regional

Transmission Organizations (RTOs) to control the transmission activity in specific areas of the continent so that power trading is able to extend over very long distances. Several regional grids have been established and are run by private boards of directors.[7]

This creation of very large transmission areas has radically transformed the nature of electricity trade between Canada and the US. Electricity exports from Canada to the US under geographically contained electricity systems depended primarily on surplus power being generated and the selling of power through long-term purchase agreements. With very large transmission systems power traders, who do not normally generate electricity, are able to buy power in one jurisdiction and sell it in another, mainly through short-term purchase agreements or on the spot-market. These changes in the market have greatly spurred the participation of private energy providers in the North American market.

The creation of any RTO requires that all utilities give up the operation and control of their transmission systems to the new entity. Each RTO is set up as a private company and no utility has a voice in its governance structures or its operations.

The major change this makes for Canadian public utilities it that it will give a private US company control over the entire electricity system. Any RTO ultimately will have the authority to set prices, enact all interchange schedules, maintain system reliability and security, and plan for future expansion of the system. While the utilities may still own the assets, that is the transmission lines and control centres, the private RTO will be able to determine the extent of new investment, its nature and, thereby, who gets the electricity.

In conjunction with the creation of RTOs, FERC also tried to establish a "seamless" marketplace for electricity throughout the continent through what it has called a standard market design. This standardization would require all markets throughout North America to follow identical structures, including the breaking up of all integrated utilities—even those in the public sector. This policy amounts to a profound and thorough redesign of the entire US electricity market and, of course, the Canadian market as well. It affects Canada because the US insists that any Canadian provinces that export electricity to the US adopt an identical system to that in the US. This demand, by the way, flies in the face of NAFTA, and if Canadian governments wanted to protect public systems of providing electricity NAFTA would be the vehicle to do this.

Many US state governments, utilities and consumer groups, most notably in the West and the South, reacted very negatively to FERC's invasion of their regulatory territory through the demand for SMD. Surprisingly, there was no resistance from any governments or consumer groups in Canada. This was such a sore point in the US that in an effort to have the Energy

Bill passed, Congress forced FERC to hold off ruling on standard market design until the end of 2006. Nevertheless, according to FERC's chairman, the way "voluntary" RTO's are proceeding means that standard market design is already taking shape. He therefore thinks that specific rulings will not actually be necessary. In addition, the new US *Energy Policy Act* of 2005 has accomplished a great deal of what was intended through SMD.

US Energy Bill

The US Energy Bill, the *Energy Policy Act* of 2005, gives FERC considerably more powers to extend its authority in a wide variety of energy issues. According to FERC's chairman Joseph Kelliher these new powers constitute "the most important change in federal electricity and gas laws since the 1930s."[9] The most significant of these changes for Canadians are the powers to oversee reliability standards, the ability to site transmission facilities and the repeal of the *Public Utility Holding Company Act.*

Electricity Reliability Organization

The EPA empowered FERC with organizing rules to govern the setting up of a private agency to establish reliability standards in North America.[10] In early 2007, FERC approved the North American Electric Reliability Corporation (NERC) as the organization that will both establish the standards and oversee grid reliability.[11] FERC itself has been given the regulatory power to ensure that the standards established by NERC are enforced. The bill specifically mentions provisions for "taking, after certification, appropriate steps to gain recognition in Canada and Mexico."[12] Making reliability a US national regulatory issue arises primarily as a result of the August 2003 blackout.[13]

The problem for Canadian jurisdictions is that these standards will govern not just the US but the entire the North American grid systems—a further incursion into Canadian regulatory authority—without explicit negotiations with Canadians. Canada and Mexico are included after decisions are made. The developments of an integrated grid system and the establishment of RTOs have implicitly integrated the systems and anyone participating on the grids will need to meet US standards. The rules will be imposed without bi-national treaty-making and will be enforced without bi-national panels, dispute settlement mechanisms or consultation. FERC will be in charge of the grid.

What is worse, reliability on the grid system will not be improved over what exists. This is because the main causes of system problems are not being addressed. According to one group of analysts, "The risk of a massive blackout is no lower today than it was in 2003—despite the passage of the Energy Policy Act of 2005, which will do nothing to enhance reliability. In fact, power system reliability may be even more at risk in the future."[14]

Repeal of the Public Utility Holding Company Act

Under the *Public Utility Holding Company Act* (PUHCA), which was established in 1978 to encourage alternative fuels for electricity generation, utility holding company systems were limited to a single region. Each utility operated much as Canadian utilities have done—that is each utility would comprise a single, integrated system. The repeal of PUHCA enables large investor-owned utilities to enter into and buy outside their core business and allows investor-owned utilities to buy other utilities without having the interconnection requirement of PUHCA.[15] The clear implication is that increased concentration in the system will occur and much larger areas will come under the control of single owners.

The intent in repealing this act is to encourage the private sector to increase its investment in the electricity infrastructure in the US. This change ultimately could effectively eliminate local control and accountability and any regulation of rates. It also means that large US-based companies are free to enter electricity markets in Canada.

Expanded Authority on Transmission

Through the *Energy Policy Act*, FERC has been given authority (similar to the pipeline authority it has had) to determine and order transmission sites. This preempts states' rights by shifting authority for transmission lines to the federal sector and FERC. This will result in fewer parties having input and control over very large sections of the grid.[16] The EPA also extends FERC's authority to order open access transmission service to entities that formally came under state jurisdiction. FERC's jurisdiction now extends to include federal, state and municipal transmitting utilities.[17] It also orders federal utilities to join RTOs. These include federal power marketing agencies and the Tennessee Valley Authority.[18]

The implications of these major changes for Canada are significant. These expanded powers for FERC will give FERC much more authority in designing the North American grid system and in regulating how it is used in the future. These actions have been taken without reference to the needs of either Canadian or Mexican markets and without consultation with other affected jurisdictions.

The US plans for electricity restructuring assume that competition will bring about abundant supplies and lower prices. In the face of the rather monumental restructuring failures that have actually occurred in the US, this approach seems more faith-based than reason would allow.[19] The weird reluctance to face up to the problems of restructuring—namely inadequate supplies, transmission congestion, price escalation and more government subsidies of private electricity development just to get more generation built can only be accounted for by a government that is firmly under the control of the private energy companies or one that follows a blind adherence to the idea that the market is always efficient—even when it clearly isn't. Of course, there is always the possibility that both factors are operating.

One of the major problems for Canada is that its energy regulatory body, the National Energy Board (NEB), is extremely weak, at least relative to the enormous powers of the US regulator, the Federal Energy Regulatory Commission (FERC).

Canada's Responses

In Canada the response of governments to US dictates has been surprisingly limp: rather than challenging US regulatory imperialism as, ironically, many US states and state public utilities are doing, Canadian governments at both the national and provincial levels are readily acquiescing to US demands. All of this is occurring with virtually no public debate, government analysis or media scrutiny.

In contrast to FERC, Canada's regulatory body, the National Energy Board (NEB), has very few powers and even those they have are not all exercised. The NEB's relocation from Ottawa to Calgary in 1991 ensures that it completely supports private energy—of any description. In a decided understatement, Doern and Gattinger note that the relocation to Calgary and the funding of the board through cost-recovery "were seen by many in the Board and in the industry as positive pro-market steps."[20] This pro-market approach of the regulator makes it not surprising that the NEB has had no role in resisting US attempts to redesign the North American system in its interests.

The NEB's mandate on electricity is restricted to the regulation of exports and the construction of facilities related to international trade. Unlike FERC, the NEB does not regulate energy within provincial boundaries, inter-provincial electricity trade or energy emissions. But in recent years, even the NEB's close monitoring of exports, which in the past included public hearings on each application for an export permit, has been replaced by blanket export permits that last for up to ten years. These appear to be given very easily and certainly without public scrutiny. The result is that export permits are frequently given to companies that do not produce any electricity at all. For example, Duke Energy, headquartered in Charlotte, North Carolina, was given a ten-year blanket export permit to export electricity from British Columbia even though it had no generating facilities. Big companies see their future in trading electricity—much in the way that Enron did—not in producing energy.

Implications for Canada

All major electricity-exporting provinces in Canada have complied in some measure with FERC orders for breaking up the integrated nature of the utilities. While provincial governments in Canada seem to realize the US is encroaching on their regulatory areas, they are cooperating with FERC to a much greater extent than many of their US counterparts.

One of the major implications for Canada in the new design for the transmission market is that it will encourage the system to expand in order to increase the export and import of power and to encourage private electricity generation. When increased access to US markets occur, as is the intention of the RTOs, all new private energy generation in both countries will have the option of selling within the province or selling in the US. This will result in domestic consumers competing with American consumers for power produced within the country, very much as they do for oil and gas. And, since the prices in the US are higher, they will get much higher here too.

New investments in cross-border transmission lines could well turn out to be very expensive for provinces in Canada. This is particularly the case considering the possibility of expanded private generation and the relatively small proportion of electricity that can now be exported through existing transmission lines. Since for the most part the wires will still be in the public sector in Canada, it very likely will be the public that will be paying for the expansion of the system—primarily to suit the requirements of the private sector and the export markets.

Developing a private electricity system within an integrated North America market creates huge problems. One is related to the relentless increase in the sheer size of the electricity markets and the distances over which electricity is transported. The electricity grids between Canada and the US serve two main purposes: one is to ensure the reliability of the system and the other is to permit trading of electricity. But the main issue in the creation of continent-wide markets is the extent that the objectives of trade itself will over-ride other significant domestic objectives of delivering electricity. These include social objectives of equity, low costs, regional development, aboriginal rights, reliability and conservation. As trading areas extend thousands of miles across the continent, efficiencies are lost, reliability of the system is compromised and meeting local needs can be superseded by the lure of large incomes from exports.

The separation of transmission from the comprehensive planning of utilities was largely responsible for increased system unreliability and was responsible for the spectacular power blackout on August 14, 2003.[21] This blackout knocked out power to over 50 million people, shut down more than 100 power-generating facilities, and closed major financial and industrial centers in the US northeast and in Ontario. But, rather than responding with a rethinking about the direction of restructuring, the US called for greater integration of the systems under the control of the US regulator, FERC.

The negative effect on planning for social needs can be stark. As transmission and generating systems expand to meet the need of private energy producers, local government and aboriginal rights tend to receive brutal treatment by governments and the private sector. In British Columbia local governments strenuously resisted the privatization of local rivers through "run-of-the-river" hydro-generation facilities[22] and enacted local legislation restricting these activities. Communities resisted these projects because of the environmental degradation involved, the negative effect it would have on local economy and the setting aside of Aboriginal land-claim issues through the de facto development of transmission lines. The government acted swiftly by passing the infamous Bill 30 in 2006.[23] This bill permanently sets aside local governments' ability to have any say in the development of private hydro electric generation facilities. It completely overrides local zoning and planning regulations and, in effect, forces Aboriginal groups and local governments to "bargain" with the private sector to have some of their needs met. In some places Aboriginal groups have been able to forestall the rampant development of transmission lines, but as private electricity initiatives escalate, the ability to withstand further encroachment on aboriginal lands will require enormous and continuous resistance.[24]

A second problem created by the restructuring of the electricity sector is the startling increase in electricity trading by corporations that do not produce electricity, but buy and sell it to take advantage of different prices in different areas of the continent. While Enron's trading needs was the ignition that brought about the North American system redesign and the new rules to facilitate traders, its initiatives have taken a life of their own long after its activities have been discredited.

A third problem related to restructuring comes from the attempts to deregulate some parts of the electricity business (generation), while retaining the monopoly aspects of other aspects (transmission and distribution). The technology of transmission has not changed its characteristic as a natural monopoly. This is mainly because the construction of a transmission system is complex, expensive and does not efficiently allow for competing transmission lines. The result is a hybrid system with a competitive market in electricity generation that encourages increased supply coupled with a limited and monopolistic transmission system. The bottlenecks that are created, then, tend to limit the expansion of the generation market and have a tendency to increase the unreliability of the system itself. It is this problem that is most crucial in overcoming the barriers that now exist to a continent-wide electricity market.

Choices for Canadians

The main issue before Canada is whether electricity systems should remain independent and controlled by Canadian governments, or subsumed within the US system. Integrating the US, Canadian and Mexican electricity markets, which is the goal of FERC, will result in prices that are established by US markets, and regulations that further the energy objectives of the US and private companies.

This does not need to happen. The North American Free Trade Agreement (NAFTA) permits both trading and investment without instituting standard market designs. There is no requirement in international law that any entity in Canada has to completely change its system in order to export into the US. This is a fundamental protection that has been retained under NAFTA.

According to the NAFTA Commission for Environmental Cooperation in its assessment of the cross-border electricity trade, provincial decisions to acquiesce to FERC demands are voluntary and the US has no right to insist on identical systems in order to trade. Under NAFTA no province in Canada is required to have exactly the same kind of organization of its market or industry as exists in the US. Canada must, however, grant "national treatment" to foreign firms. What this means is that as long as a government treats private domestic and foreign firms in the same way, it is not contravening NAFTA.

In order to use the protections of NAFTA, Canada would need to have a national government that is pro-active in protecting Canada's electricity interests. The absence of a strong Canadian presence becomes glaringly evident in the negotiations with the US over market design and transmission organizations. Each province is basically on its own in determining its relationship with the US. This is unfortunate because the impression FERC projects in its drive to control the entire North American electricity industry is that Canadian electricity systems will have to mirror developments in the US in order to have access to the American market.

Options

Canada has a strong legal position to maintain public provision of electricity. Among some important actions that would need to take place would be the following:

- Electricity, currently a provincial issue, needs to be treated as a national one. The increased internationalization of US regulatory design requires a strong national voice in negotiations with the US. Electricity is no longer confined by provincial boundaries and in the face of US regulatory imperialism Canada can justify taking action in the design of future markets.

- The federal government should adamantly resist the development of a "seamless North American electricity market," which requires the complete integration of provincial electricity systems with the US system.
- Canada should use the protections in NAFTA to allow public utilities and provincial governments to maintain integrated utilities in the public sector.
- The federal government should encourage greater integration of the Canadian electricity sector. Currently each province has greater ties with the US than it does with other provinces. This is partly a result of the regulatory vacuum at the national level. With the need for market reliability and for new investments in electricity generation increased inter-provincial planning would make a lot of sense. The US has a regulator that deals with national and international issues. It is time that we in Canada had one too.
- Canadian public entities should maintain control of transmissions systems and not surrender any part of these systems to foreign controlled organizations (like the Regional Transmission Organizations).
- All transmission systems should be owned and operated by public entities.
- Canada should prohibit private exports of electricity and private power trading.
- The federal government should have a strong regulatory role in electricity related environmental issues. Related to this it should do the following:

  - Institute a nation-wide electricity conservation program.
  - Require provincial governments to conduct system-wide assessment of all private power projects for their cumulative effects (currently they are assessed individually).
  - Establish an institute for environmentally responsible energy development and invest heavily in green energy development in the public sector.
  - Ensure that democratic principles are adhered to in decisions about future energy production including open public hearings on new energy projects, intervener funding for popular sector and first nations groups at the regulators' hearings, and placing a priority on first nations issues in all new energy projects.

Conclusion

It is time for Canada to recognize that it needs a clearly defined energy policy. So far there is no sense that energy security, which is much on the minds of the US and other countries, figures at all in Canadian government policy. Canada gave up control over oil and gas and appears not to rethink this even with regard to the treatment of new reserves and new exploration. In this Canada is distinct from most of the world, where the largest oil companies and the largest reserves are owned by nation states.

If Canada begins to treat electricity as it has oil and gas, we can expect very similar results and loss of control of a crucial resource. Canada is free to makes it own decisions and need not adopt the US strategy for a deep integration of electricity markets. NAFTA allows both trading and investment across borders without having to establish identical market structures. But in order to pursue this route, at the very least, the Canadian government would need to have a plan and become pro-active in protecting the country's interests.

Notes

1. Dick Cheney, et al., *National Energy Policy: Report of the National Energy Policy Development Group* (Washington, DC: US Government Printing Office, 2001).
2. Vertical integration means that generation, transmission and distribution systems are owned and managed by the same corporation and all parts of the electricity system are planned in coordination.
3. David M. Newbury, *Privatization, Restructuring, and Regulation of Network Utilities* (Cambridge, Mass: MIT Press, 2001), p. 259.
4. Electricity has many properties of a "natural monopoly" that made breaking up the integrated systems difficult. Ultimately electricity deregulation succeeded largely as a result of the efforts of Enron. Enron was much less interested in generating electricity than in trading it from one jurisdiction to another.
5. Marjorie Griffin Cohen, "International Forces Driving Electricity Deregulation in the Semi-Periphery: The Case of Canada," in Marjorie Griffin Cohen and Stephen Clarkson, eds., *Governing Under Stress: Middle Powers and the Challenge of Globalization* (London and New York: Zed Books, 2004).
6. John Kwoka, "Restructuring the US Electric Power Sector," American Public Power Association, November 2006, http://www.appanet.org/files/PDFs/RestructuringStudy/Kwoka1.pdf, June 6, 2007; G. Bruce Doern and Monica Gattinger, *Power Switch: Energy Regulatory Governance in the Twenty-First Century* (Toronto: University of Toronto Press, 2003).
7. Tyson Slocum, "The Failure of Electricity Deregulation: History, Status and Needed Reforms," Public Citizen's Energy Program, http://www.citzen.org, accessed June 6, 2007.
8. Marjorie Griffin Cohen, "Imperialist Regulation: US Electricity Market Designs and Their Problems for Canada and Mexico," in Ricardo Grinspun and Yasmine Shamsie, eds., *Whose Canada? Continental Integration, Fortress North America and the Corporate Agenda* (Montreal: McGill-Queen's University Press, 2007).
9. Joseph Kelliher, Interview on "Money and Politics," BLTV, August 17, 2005. http://www.multivision.com, accessed October 1, 2005.

10. US Government, *Energy Policy Act* (2005), Title XII, Subtitle A, Section 1211, Section 215, (b) (3).
11. Federal Energy Regulatory Commission, "Commission Accepts NERC's Regional Reliability Entity Delegation Agreements," Docket Nos. RR06-1-004, April 19, 2007.
12. EPA, 2005, Title XII, Subtitle A, Section 1211, Section 215, (c) (B) (v).
13. This massive blackout originated in the US, despite initial efforts to shift the blame to Ontario.
14. Jack Casazza, Frank Delea and George Loehr (2005). "Contributions of the Restructuring of the Electric Power Industry to the August 14, 2003 Blackout," Paper presented at technical workshop on Competition and Reliability in North American Energy Markets, Toronto, September 28, 2005.
15. EPA, 2005, Title XII, Subtitle G, Section 1273.
16. EPA, 2005, Title XII, Subtitle C, Section 1233; (1) (A).
17. EPA, 2005, Title XII, Subtitle A, Section 1211; Section 215, (b) (1).
18. EPA, 2005; Title XII, Subtitle F, Section 1266; (B) (5).
19. Slocum, "The Failure of Electricity Deregulation."
20. Doern and Gattinger, *Power Switch*, p. 103.
21. Casazza, Delea and Loehr, "Contributions."
22. This term implies that rivers remain relatively untouched through the electricity generation process, but this is far from the reality. Water is diverted from the river, often for many kilometers and sometimes through mountains, leaving as little at 10% of the water flow in the river.
23. John Calvert, *Sticker Shock: The Impending Cost of BC Hydro's Shift to Private Power Developers* (Vancouver: Canadian Centre for Policy Alternatives, 2007)
24. For information about Aboriginal rights and transmission lines in Manitoba see Cy Gonick, "Gary Doer's Manitoba," *Canadian Dimension*, July/August 2007.

Climate Change and Energy Security for Canadians

by Gordon Laxer

In Washington and London, Prime Minister Stephen Harper bragged that Canada is a new "emerging energy superpower," and a secure source of almost limitless energy resources.[1] Is Canada really an energy superpower? A true superpower is one that can influence events by projecting economic, military, political and cultural power on a world scale.

Critics argue that, on the contrary, Canada is a US-resource satellite, or neo-colony.[2] Satellites are formally independent countries, dominated through deep ideological allegiance and economic dependence on a more powerful country.

Which view is closer to the mark? On the superpower side, Canada has one of the world's greatest supplies of oil, more than in Saudi Arabia. Ninety-eight percent of it is in Alberta's tar sands. Canada has large supplies of coal, uranium and hydropower too, but not yet many alternative power sources. On the other hand natural gas and conventional oil are running out, with fewer than ten years "proven" supplies of each.[3] On balance then, Canada has the potential to be an energy superpower. But, Canada's tar sands face several almost insurmountable obstacles that will likely limit production levels.

To make the goo run, the tar must be heated and hydrogen added. Natural gas is currently the main source of both. If Canada gets tough on carbon emissions, it will no longer permit unlimited tar-sand expansion. Greenhouse-gas intensity in the tar sands is almost triple that of conventional oil. Alberta now emits more greenhouse gases than Ontario, despite having only 26 percent of Ontario's population.[4] Nuclear energy is being touted as a carbon-neutral replacement for natural gas to heat the tar sands, but is fraught with environmental dangers. While nuclear-power generation is carbon-neutral, refining uranium can emit a lot of carbon dioxide. There are no long-term, safe methods to dispose of spent uranium, and nuclear accidents can happen, with devastating results.

Heating the tar sands is only one of the environmental obstacles that limit Canada's oil output. The tarry sands require two to five barrels of water to produce one barrel of oil. Alberta's northern river flows are dwindling under the burden.[5] Finally, construction and extraction costs are currently very high. On the other hand, Canada is running out of cheaper and less environmentally damaging conventional oil, with fewer than ten years of "proven" supplies remaining.

But environmental obstacles are not the only thing blocking Canada from becoming an energy superpower. The US empire's obsession with gaining control over the world's energy supplies by any means and the sway of oil transnationals in Washington and Ottawa make assertions of Canadian energy superpowerdom ludicrous. A day after the Harper Conservatives won the 2006 election, officials from Natural Resources Canada met in Houston with their US counterparts.6 The latter urged Canada to increase tar-sands production five-fold and export most of it to the United States.

Superpowers are supposed to influence world events. What kind of superpower lets foreign corporations dominate7 their resources, and practically gives away tar-sand oil, its biggest resource, at as low as twenty-five cents per barrel to giant transnationals like Exxon? What kind of superpower gives up the right to set its own domestic oil and gas prices, block foreign ownership of the energy industry, or limit exports, so it can cut greenhouse gases?

Paradigm Shift

The ground has shifted dramatically since the 1990s. Gone is talk about a borderless world, cheap oil, cavalier denials of climate change and transnational oil corporations ruling the roost. Instead, we are witnessing "security trumping trade," disastrous wars for oil and rising energy prices. We also see energy transnationals losing their grip and government energy-security strategies partly supplanting open markets.

In some ways, it's a return to the past. The power to control the world's oil has recently shifted sharply from the US and the transnationals to oil-producing countries. State-owned companies dominate again, and now hold 77 percent of the world's oil supplies. Long-term state-to-state contracts are replacing spot markets. This means that Arab countries could for the first time in twenty-five years effectively boycott the US, as they did with considerable psychological effect8 in 1973–4. What is new is that this power shift coincides with political elites suddenly taking climate change seriously and security issues superseding open trade and investment.

In this chapter, I show how climate change issues connect with those around energy security for Canadians and popular national control, demanding a paradigm shift.

While many other countries are grappling with today's new challenges, political debate in Canada continues as if the world is unchanged. International treaties and other neoliberal obstacles left over from the late twentieth century currently hold Canada in the old paradigm. In the North

American Free Trade Agreement (NAFTA), Canada foolishly guaranteed the US unlimited access to Canadian energy, even if Canadians were to run short during an Arctic cold front. Until Canada grasps "energy independence," and a new "national energy policy," as US leaders call their plans, Canada cannot seriously address climate change. To do the latter, Canada would have to regain control over energy exports, ownership and production levels. Canada needs a new energy strategy.

Proportionality is the major problem. A technical term that conceals how much Canada's hands are tied, proportionality was part of the 1989 Canada-US Free Trade Agreement and continues in NAFTA. Proportionality forces Canada to export the same share of energy—63 percent of Canadian oil production and 56 percent of natural gas9—as in the past three years.10

Mandatory proportionality currently reigns, but need not continue to. Canada can demand a Mexican exemption. Mexico is a NAFTA member, but wisely exempted itself from proportionality's compulsions.[11] Canada can and should demand a similar exemption. If the US refused Canada's demand, as President Clinton did in 1993,[12] Canada can exit NAFTA by giving six months notice.[13]

Readers may wonder how Canada got stuck with proportionality, which is so contrary to Canadians' interests. The answer is political power. In the 1980s, proportionality was promoted by oil and gas corporations as a way to gain access to US markets during an oil and gas slump in Canada. The corporations aimed to gain quick profits by exporting Canadian energy in its raw forms. Naturally, US subsidiaries attempted to secure Canadian energy for use by their parent company. Domestic corporations saw greater profits in exporting to the large US market, than confining themselves to Canada.

No one should be surprised when corporations act in their self-interest, but citizens expect their governments to protect the public good. This did not happen. Conservative governments in Alberta and Ottawa worked very closely with the oil corporations to overturn the far-sighted energy independence policies of the 1970s and early 1980s. Promoted by the Conservative governments, the *proportionality* clause was inserted into the free-trade agreement with the United States to ensure that future Canadian governments would not reverse it, and thereby dim prospects for huge, future oil profits. Former Alberta Premier Peter Lougheed declared, "The biggest plus of this [Free Trade] agreement is that it could preclude a federal government from bringing in a National Energy Program ever again."[14]

They called the Agreement "free trade" to appeal to Canadians, but the international deal was mainly about guaranteeing the US unlimited access to Canadian energy,[15] and recognizing the rights of unelected corporations above those of elected governments.[16]

Continental vs Canadian Energy Strategies

Harold Innis, the great Canadian historian, wryly commented on the standard story of Canada evolving from "colony to nation." Instead, "Canada moved from colony to nation to colony," he asserted. The implication was that Canada enjoyed a brief period of independence as it shifted from "British imperialism to American imperialism."[17] Today, the question is whether Canada can move to independent "nation" status again, particularly regarding energy. To understand how Canadians can retake control over their energy future and cut greenhouse gases, we need a brief look at the tug of war between national and continental energy policies.

By the 1950s, it was evident that bountiful American resources were running out, particularly in energy. Consequently, America's corporate and political elites increasingly cast covetous eyes on Canada's plentiful supply, as the most secure and receptive source for their insatiable appetites. Two roads were possible: (1) foreign ownership and (2) shaping Canadian policies. Standard Oil's[18] takeover of Canadian-owned Imperial Oil in 1899 began a trend to foreign domination of Canadian oil and gas that reached 90 percent by 1960.[19]

Influencing Canadian policies was the other road to controlling Canada's energy resources. US petroleum corporations have pushed Washington to pressure Canada into exporting energy on a "US-first" basis. American corporations have also influenced Canada from the inside. As English-speaking managers of foreign subsidiaries, they could "pass" as part of Canada's corporate community. In 1965, George Grant portrayed foreign corporate sway inside Canada this way: "foreign capital is able to determine possible governments by incarnating itself as an indigenous ruling class."[20]

By the 1960s, Canada was copying American energy policies. Because US oil supplies cost more than Middle Eastern and Venezuelan sources, the US imposed quotas to protect most of the US market from cheap imports. Western Canadian oil, high-cost like US oil, was treated as if it were American oil and exempted from US import quotas. Ontario and the western provinces were reserved for expensive western Canadian oil. From 1961 to 1973, Canada's National Oil Policy (NOP), paralleled US policies. Ontarians subsidized western Canadian oil, while cheaper imports supplied Atlantic Canada and Quebec. This pattern made sense for US transnationals, but left Canada with a continental energy policy which served the interests of the US more than Canada.[21]

The 1970s was dominated by international oil crises and dramatic policy shifts. The Organization of Petroleum Exporting Countries (OPEC), a grouping of mainly Middle-Eastern, oil exporting countries began asserting national and government control of their oil resources. As owners of their resources, they demanded and got higher royalties and helped drive up the international oil price by more than ten times over a very short period. The energy world was turned upside down, and Canada responded by abandoning the continental, pro-corporate policies long advocated by big foreign oil. Canada boldly began to forge a Canada-first strategy and challenged control by the transnationals.

Canada briefly gained energy independence, demonstrating that energy sovereignty was an option. A new Canadian nationalism strengthened at the time of Canada's 1967 centennial and Expo 67, Montreal's wildly successful world exposition. It was a time when many Canadians and those in other countries saw the United States in increasingly negative terms—the war against Vietnam, race riots and gun culture. Canadians embraced economic nationalism and, like many people abroad, desired to control their own resources.

Pushed by the political left and citizens' movements through the 1970s, the Trudeau government adopted Canada-first energy policies. These culminated in the National Energy Program (NEP) in 1980. The Canadianization goals of the NEP were wildly popular, hitting an 84 percent approval rating.[22] The goal was to end foreign transnationals' domination over Canadian oil and natural gas by rapidly expanding PetroCanada, the government's oil company, and through promoting private Canadian ownership.

The Canada-first policies of the 1970s and the NEP reduced oil exports to the US and sent western Canadian supplies east to partly displace oil imports. Canada continued to export oil though and charged US buyers the world price, which was higher than Canada's domestic price. The differential between the Canadian and the world price was captured by the federal government. It used the funds to lower oil prices for those eastern Canadians still consuming foreign oil. In this way, although Canada did not supply domestic oil to all Canadians, they all paid the same, low price, wherever they lived. As one can imagine, this was a very popular policy in eastern and central Canada.

Keeping Quebec in Canada was a major goal of the government's bold energy-independence agenda. At a time when the Parti Québécois had been elected to its first term and held its first sovereignty referendum, the Canada-first energy policies were presented to Quebecers as a reason to stay in Canada. Quebec has no oil or natural gas of its own. Low Canadian oil prices and a share of energy revenues were attractive to Quebecers.

While popular in eastern Canada, the NEP was greatly resented by big oil and the US government. After a first burst of support for its Canadianizing policies, the NEP caused an explosive reaction in Alberta, home to Canada's oil and gas industry. Provinces, not the federal government, own crown resources. The NEP was seen widely as a federal attack on Albertans' rightful share of their energy wealth. They were partly right. Alberta's counterattack was based in the foreign-controlled oil corporations, but was also genuinely populist. Hence the popularity of a bumper sticker which read "let the Eastern bastards freeze in the dark." Alberta's government substantially cut oil supplies to eastern Canada and economic civil war ensued.[23]

Timing was unfortunate for the NEP. World oil prices crashed in 1982, which led to a widespread debt crisis in the global South. Although all oil producing regions in the world suffered a devastating crash after enjoying boom times, most Albertans blamed the NEP for Alberta's long economic crash of the 1980s. They were helped along in this view of the cause of their misfortune by a hysterical corporate media and Alberta government. The stage was set for overturning the National Energy Program and inserting the proportionality clause to ensure it would never rise again. The old NEP was gone, to be replaced by a new NEP—"No Energy Policy," which still reigns today.

A Canadian Energy and Conservation Strategy

That was the context in which Canada gave up control over domestic energy resources and many of the necessary means to substantially limit greenhouse gases. But Canada did not give up control forever. Canada can, indeed it must, develop a new strategy and paradigm to deal with the issues of the twenty-first century—the threat of climate change and energy insecurity. What would such a program entail?[24]

- Impose an immediate moratorium on new tar-sands projects. The tar sands are the single largest contributor to greenhouse gases in Canada.
- Get a Mexican exemption on compulsory energy exports (proportionality) or give six-months notice and exit NAFTA. It will be difficult to convince Canadians to drastically cut energy use if most of the energy they save is exported to the US under proportionality.
- End oil imports to eastern Canada. Don't let the "eastern bastards" freeze in the dark. Reduce, not end, oil exports to the US to do this.
- Reinstitute the twenty-five years of "proven" supply of oil and gas for Canadians before export licences can be issued.
- Build enough oil pipeline capacity to Montreal to ship western oil to eastern Canada. Redirect Nova Scotia's natural gas and Newfoundland's oil from exports to the New England states to meet Atlantic Canadians' needs.
- Build a national electricity grid for power sharing across Canada.
- Then, instead of building more tar-sands plants at $10 to $12 billion

per project, provincial and federal governments should adopt policies to reduce energy usage and carbon consumption for a fraction of the cost.

- One hundred percent of the economic rents (excess profits) on existing oil and gas should go to the resource owners—citizens of energy-producing provinces and First Nations. Norway gets three to six times as much economic rent per barrel of oil as Alberta. The latter is giving away billions in "unearned" or "'excess" profits to Exxon-Mobil, Shell and other oil transnationals.

- Use the extra resource revenues collected to fund *just transitions* to a post-carbon society. Just transitions include guaranteeing alternative jobs for workers in fossil fuel industries and making sure that higher prices on fossil fuels do not block low-income people from gaining adequate access to home heating and transportation in a post-carbon society. It also means that all sectors, especially industry, and rich people, cut their disproportionate carbon emissions. They will not be permitted to shirk their responsibilities by paying someone else to cut emissions in their place. No one and no corporation should have more right to foul the atmosphere than anyone else. Thus the rich must reduce their emissions to levels of the poor.

- Follow Canadian public opinion and nationalize or provincialize part or all of the oil and gas industry. This would ensure continued Canadian control. Private Canadian owners are often targeted for takeover by foreign owners. Profit-driven corporations are limited to pursuing the bottom line, but crown ownership allows corporations to fulfil public-policy goals of energy conservation and upgrade resources in Canada before exports. A Leger poll in 2005 showed that 51 percent of Canadians with an opinion supported nationalizing the oil corporations, including 60 percent of the young.

Conclusion

With the mounting evidence of a looming climate-change catastrophe, most Canadians support urgent action. To meet its international Kyoto targets and then go far beyond them, Canada needs to reduce oil and gas exports and drastically cut energy consumption. Canada should forget about fantasies of energy superpowerdom and instead become a world leader in making a just transition to a post-carbon society.

Notes

1. Jane Taber, "PM brands Canada an 'energy superpower,'" *Globe and Mail*, July 15, 2006, http://www.theglobeandmail.com/servlet/story/RTGAM.20060715.
2. Michael Vickerman, "Draining Canada first," http://www.energybulletin.net/21899.html, ac-

cessed November 2, 2006.

3. Statistics Canada, *Human Activity and the Environment. Energy in Canada,* 2004, Table 1.8, p. 8.

4. *Globe and Mail,* "Canada's greenhouse-gas emissions increase," November 28, 2005, http://www.theglobeandmail.com/servlet/story/RTGAM.20051128.wxemissions28/EmailBNStory/National/.

6. Dan Woynillowicz, *Oil Sands Fever. The Environmental Implications of Canada's Oil Sands Rush* (Drayton Valley, AB: Pembina Institute, 2005).

6. CBC, "U.S. urges 'fivefold expansion' in Alberta oilsands production," www.cbc.ca/canada/story/2007/01/17/oil-sands.html, accessed January 18, 2007.

7. With respect to the assets of Canada's oil and gas industry, 49 percent are foreign-controlled. See Statistics Canada, *Canadian Economic Observer,* catalogue number 11-010, March 2006, p. 315.

8. There is controversy about the boycott's economic effectiveness. Contemporary newspapers reported oil tankers parked off North American shores, waiting for prices to rise before landing.

9. These ratios are the level of exports over total Canadian production. NAFTA calculates proportionality by relating exports to total supply—Canadian production plus imports plus inventory draw-downs. See John Dillon, "How NAFTA Limits our Energy Options," unpublished paper, September 2006.

10. NAFTA, Article 2102, allows Canada to override proportionality, but only temporarily during war or other international emergencies. See Canada, North American Free Trade Agreement, 1994, http://www.dfait-maeci.gc.ca/nafta-alena/agree-en.asp#PartI.

11. NAFTA, Article 605.

12. Julian Beltrame, "U.S. says no to revising energy deal," *Ottawa Citizen,* November 20, 1993, p. A1.

13. NAFTA, Article 2205.

14. Peter Lougheed, "The Rape of the National Energy Program Will Never Happen Again," in Earle Gray, ed., *Free Trade, Free Canada* (Woodville, ON: Canadian Speeches, 1988).

15. NAFTA, Chapter 6.

16. NAFTA, Chapter 11.

17. Harold Innis, "Great Britain, The United States and Canada," in Mary Q. Innis, ed., *Essays in Canadian Economic History: Harold Innis* (Toronto: University of Toronto Press, 1956), p. 405.

18. John D. Rockefeller's Standard Oil was the major forerunner of today's Exxon-Mobil.

19. James Laxer, *Oil and Gas. Ottawa, the Provinces and the Petroleum Industry* (Toronto: Lorimer, 1983), p. 7.

20. George Grant, *Lament for a Nation: The Defeat of Canadian Nationalism* (Toronto: McClelland and Stewart, 1965).

21. Laxer, *Oil and Gas,* p. 8.

22. *Daily Oil Bulletin,* September 15, 1981.

23. G. Bruce Doern and Glen Toner, *The Politics of Energy. The Development and Implementation of the NEP* (Toronto: Methuen, 1985), pp. 266–75.

24. Dave Thompson, Diana Gibson and Gordon Laxer, *Towards an Energy Security Strategy for Canada* (Edmonton: Parkland Institute, 2005).

Peak Oil and the Future of Agriculture

Has Cuba Shown the Way?

by Paul Phillips

Agriculture, as it is now practised in North America and exported via the World Bank and the International Monetary Fund to developing countries as a workable model, is unsustainable. This high input, industrial agriculture, "green revolution" model, which has been globally dominant for the last half-century, allowing world population to double in that period, is about to run out of energy and water.

As the world's demand for energy outgrows new supplies—something that will happen whether or not production of oil has peaked or is about to peak—the rising price of oil and gas will ensure that farmers will not be able to afford to produce food at prices that consumers can afford to pay.

Furthermore, the water that is required for the industrial model of agriculture is drying up, in part due to global warming, in part due to overuse of supplies drawn from the world's major aquifers. Just this past year, a report from the University of Alberta predicted a new period of drought for the prairie region as bad or worse than that of the Dirty Thirties. What is more, the melting of the Rocky Mountain glaciers threatens the supply of water to southern Alberta and its huge irrigation system. In the US at present, 85 percent of all water usage is taken by agriculture. As a result, many river systems are virtually running dry before they reach the ocean. If the wars of the late twentieth century were about oil, the wars of the twenty-first century will be about water.

The Rising Cost of Energy

The most immediate threat to our food supply, however, is the rising cost and increasingly insecure supply of energy—in particular oil, but also natural gas. The threat to the food supply lies in the relation between agriculture and energy.

Industrial agriculture requires ten calories of fuel for every calorie of food produced—and this does not include the energy used in processing, packaging, and shipping the final product to market. In North America, we consume 40 percent more energy than we receive from the sun—which is ultimately the only source of sustainable energy. Most of that 40 percent comes from the solar power stored up over millions of years in the form of fossil fuels. It took millions of years to produce the oil and gas but only a little over a century to consume around half of the world's reserves of oil and of North American supplies of gas. What this means is that if the world's population is to be fed using only the sustainable part of solar energy (sun, hydro and wind) the world can only support a population of a fraction the size of the current population. Science writer Dale Allen Pfeiffer puts it very starkly:

> For sustainability, global population will have to be reduced from the current 6.32 billion people to 2 billion—a reduction of 68 percent or over two-thirds. The end of this decade could see spiraling food prices without relief. And the coming decade could see massive starvation on a global level such as never experienced before by the human race.[1]

This direct impact of peaking supplies of oil and gas on food supplies is compounded by the indirect impacts of peak oil and global warming. Even US president George Bush recently admitted that the United States is "addicted to oil." As James Kunstler writes:

> The American way of life—which is now virtually synonymous with suburbia—can run only on reliable supplies of dependably cheap oil and gas. Even mild to moderate deviations in either price or supply will crush our economy and make the logistics of daily life impossible.[2]

A False Response to a Genuine Crisis

The reaction to the rising price and increasing insecurity of fossil-fuel supplies has been the call for alternative sources of energy. In other words, there has not been a major call for the only solution that can really address the problem: conservation and the promotion of alternative sources of transportation; or, alternately, the reorganization of our economy so as to reduce the need for fossil fuels.

This call for alternative energy has been translated into policies promoting the use of biofuels for cars and trucks—in particular, ethanol and biodiesel. The problem is that both of these alternative fuels require food stocks to produce—food stocks that otherwise would be available to feed people.

After years of denial, the sudden realization by North American governments that global warming is actually happening—and that the potential cost of climate change could be catastrophic—has only accelerated the call for the replacement of fossil fuels for transportation with biofuels. Not only that, but moves to replace petroleum-based, plastic food take-out containers and garbage bags with biodegradable vegetable-based products made from corn, potatoes, sugar cane and other agricultural raw materials has exacerbated the increased demand for edible agricultural products for non-food uses.

Environmentalists have generally been strong supporters of the shift to biofuels and bio feedstocks to replace environmentally harmful and increasingly expensive fossil fuels. They have applauded the Canadian and American governments when they mandated that gasoline for motor fuel must include a certain percentage of ethanol. The reasons are transparent. Biofuels and agricultural feedstocks are not only renewable resources, but they are carbon neutral. That is, as they grow they take carbon dioxide out of the air. Then when they are burned as fuel or otherwise decompose, they release back into the atmosphere only the carbon that they have previously absorbed. In that sense, they are far superior to fossil fuels as energy stocks or as industrial feedstocks. This is because fossil fuel feedstocks involve extracting carbon sequestered underground in the form of coal, gas and oil and converting it into carbon dioxide and other warming and polluting emissions to be released into the atmosphere.

The Drawbacks to Alternative Energy

However, there are two major flaws to solving the problems of oil prices and climate change through biofuels. One is the current industrial model of agriculture used to grow the crops needed for the production of ethanol and biodiesel. This model depends heavily on fossil fuels. The crucial measure here is known as EROEI (energy return on energy invested). The EROEI of ethanol production from corn is reportedly in the range of 1 to 1.2. In other words, the energy content of the ethanol produced is little more than the fossil-fuel energy used in its production. What this means is that the production of ethanol consumes almost as much fossil fuels as it replaces! In this respect, biodiesel is considerably more attractive with an EROEI of 3.2 from soybeans and an energy content of over twice that of the equivalent amount of ethanol.[3]

Furthermore, the corn used for ethanol production must come from crops now grown for food consumption. For instance, critics estimate that to reach the 5 percent ethanol target for retail gasoline set by President Bush will require over one-half of the entire US corn crop. It will also substantially drive up the price of corn used both for human food products (including the most consumed sweetener in processed food, corn syrup) and for feeding livestock. In turn, this will drive up the retail prices of meats and processed foods and other feed grains. Producing enough rapeseed oil to produce sufficient biodiesel to replace petroleum fuels, for example, would take something like three-quarters of the American landmass. And it has even been estimated that to produce sufficient ethanol from corn to replace all the oil used in transportation in the US would require the entire landmass of the country—leaving no land on which people could live or grow food for human consumption.

Since it imports from the US much of the grain required for feeding its livestock, Mexico stands to be a big loser as the price of US corn rises. Rising prices of imported corn will also raise the price of the domestic white corn utilized in making the staple of the Mexican diet, tortillas. This will have an overall negative impact, particularly on the living standards of the poor. Similarly, utilizing oilseeds or other oil producing plants (e.g., palm oil) to provide biodiesel will produce competition for the food required for human consumption. As well, clear cutting of forests to plant oil palm in countries such as the Philippines and Indonesia threatens deforestation of many millions of acres, with the consequent increase in global-warming gas (GWG) emissions, as well as endangering a number of animal and plant species.[4]

It is possible that new technological processes will allow non-food, carbon-neutral feedstocks such as cellulose to be utilized for ethanol production; or that biodiesel can be efficiently extracted from oil-rich algae grown in waste-water ponds at water-treatment plants. This would reduce, but not eliminate, the competition for food. However, it is not possible to have a sustainable agriculture produce the amount of fuels required for our current way of life.

As James Kunstler forcefully noted on his blog for February 5, 2007:

> This obsession with keeping the cars running at all costs could really prove fatal. It is especially unhelpful that so many self-proclaimed "greens" and political "progressives" are hung up on this monomaniacal theme. Get this: the cars are not part of the solution (whether they run on fossil fuels, vodka, used frymax[™] oil, or cow shit). They are at the heart of the problem. And trying to salvage the entire Happy Motoring system by shifting it from gasoline to other fuels will only make things much worse.[5]

The fact is that the much maligned Robert Malthus was correct: the history of the industrial age is the history of first eating guana, peat and coal and then eating oil and other fossil fuels and solar energy stored in the form of oil. Natural gas is the feedstock for commercial fertilizers, while petroleum is not only the source of fuel for agricultural equipment, tractors, trucks and irrigation systems but also the feedstock for pesticides, herbicides and insecticides. In North America, industrial agriculture, excluding transport and processing, directly consumes over a sixth of all energy consumption. Industrial agriculture is addicted to oil and gas.

Kunstler is equally blunt on the future of agriculture in America:

> We have to produce food differently. The ADM/Monsanto/Cargill model of industrial agribusiness is heading toward its Waterloo. As oil and gas deplete, we will be left with sterile soils and farming organized at an unworkable scale.[6]

What do we eat when the petroleum production peaks and supplies begin to dwindle? What happens when we begin to run out of stored up solar energy—the challenge of peak oil—or even if just the price of oil rises precipitously as demand rises faster than supply? What happens to agriculture?

Cuba: After the Oil Runs Out

Well, we happen to have an example of exactly what can happen, a case history that compresses the peak-oil scenario into a few years instead of a decade or so. This is the case of Cuba after the disintegration of the Soviet Union.[7]

After the revolution Cuba took its cues, along with the subsidies, from the Soviet Union. Agriculture was reorganized on the industrial agriculture model: large scale, mechanized collective farms, mono-cropping emphasizing exports of cash crops, heavy use of fertilizers, pesticides and herbicides. By the late 1980s, Cuban agriculture had become highly reliant on Soviet supplies of cheap fuels and petrochemicals, even more so than its American counterpart. Micheline Sheehy Skeffington writes:

> From 1960 to 1989 the main exports were all cash crops—sugar, coffee, tobacco and citrus fruits. The crops were intensively farmed on large farm collectives throughout the country. But these were largely for export and up until 1989, 55% of food consumed in Cuba was Soviet-subsidised imports. Even animal feed was 97% imported, largely maize and soya beans.[8]

Cuba boasted the highest producing dairy cow in the world. An import from Canada, White Udder was fed on imported grain since Cuban grass would not sustain such high producing animals. In other words, the Cuban dairy and cattle industry was dependent on imported grains. But Cuba was also dependent on imported and subsidized Soviet oil. By the 1980s, Cuba had more tractors per hectare than California and had become dependent on Soviet subsidies of energy, fertilizers, grains and agricultural chemicals and on Soviet markets for its major agricultural exports. But this model was unsustainable.

The fall of the Soviet Union, the subsequent collapse of Cuban industrial agriculture and Soviet markets for Cuban sugar and other agricultural exports and the continuing US boycott introduced Cuba to its "special period" when Cuban GDP fell 85 percent and food supplies collapsed. The average weight of Cubans fell twenty pounds and malnutrition, especially among children, became prevalent. The average Cuban went from 3000 calories per day to 1900—equivalent of skipping one meal a day.[9]

Cuba was faced with a stark choice: capitulate to American imperialism or revolutionize its agriculture. It chose conversion of Cuban agriculture "from large scale, oil-intensive, chemical-industrial production, to small scale, local, organic agriculture."[10] This involved a number of revolutionary agricultural reforms. One was a return to animal (oxen) power which was more efficient on the smaller scale plots—particularly after the 1993 decision to break up the large state-owned, Soviet-style farms and distribute land to the *compesinos*. Some 200,000 oxen were trained to plow.

Secondly, and perhaps most important, local, renewable production, particularly in the urban areas was promoted. In 2004, for example, some 50 to 80 percent of Havana's food was produced inside the city limits, perhaps as much as 90 percent in and around Havana amounting to some 300,000 tons of produce. Rooftop gardens, urban gardens and "organoponics" (local organic gardens) have become major producers. There are some 200 in Havana, with 100,000 small- to medium-sized urban gardens countrywide. Officially, within Cuba's urban centres, there are now 2600 large-scale organic gardens, 3600 smaller, intensive gardens and 93,948 small parcels by families for their own use. Pesticides and insecticides are also prohibited. These urban gardens also absorb greenhouse gasses and improve the urban air quality, attract bees that produce honey and also produce herbs with "medicinal properties" used to treat cancer, colds and diabetes. Naturally, buying local, fresh, seasonal food drastically reduces the energy required to process and transport food to market.

What is more, Fidel Castro recently admitted that Cuba's campaign to produce export sugar as its agricultural policy centrepiece was misguided.

The third major transformation of Cuban agriculture was the development of organic or near-organic farming. Organic farming now constitutes approximately 60 percent of non-sugar farmland in the country. Given the crisis sparked by the Soviet collapse, Cuban scientists were propelled to investigate and develop biological pest control and soil-fertility enhancement. Plant fertilizers, intercropping and biological pest and weed controls have been developed. So too has vermicompost (worm castings), which is ten times as effective as cow manure on a weight basis. In addition, a national program of fruit-tree planting was instigated.

The result of all these developments has been a gradual return to the provision of an adequate and productive agricultural industry sufficient to maintain a sustainable, subsistence food supply. Moreover, because most of the food raised in Cuba is organic, the country has probably the healthiest food supply in the world.

Nonetheless, though Cubans have gotten back their third meal a day and have adequate calories to maintain a healthy diet, they still lack desired supplies of meat and milk. This is because the Cuban climate and soil does not easily support the raising of livestock. White Udder's descendants unfortunately succumbed to the native vegetation of Cuba. What this tells us is that the industrial model of high-input, GMO agriculture is no answer to the growing agricultural and world population crisis, particularly with respect to animal protein and dairy products.

This point is highlighted in a recent report from the UN's Food and Agricultural Organization (FAO)[11] In the first half of the twenty-first century, rising population and incomes, in particular in countries like China and India, are expected to result in a doubling of the demand for meat and dairy products. However, such increases are not sustainable in a world of declining oil and water supplies and global-warming climate change. As the FAO report points out, the livestock sector already produces more human-related GWG emissions than does transportation: 9 percent of carbon dioxide emissions, 65 percent of nitrous oxide emissions (a GWG 296 times more powerful than carbon dioxide), and 65 percent of human-activity produced methane (32 times as powerful as carbon dioxide). Furthermore, the livestock sector produces 64 percent of ammonia, which contributes to acid rain.

And this is not the end the contribution by livestock to global warming. Livestock now cover 30 percent of the earth's surface, while 33 percent of arable land is utilized for grain production as animal feed. Twenty percent of pastureland has been degraded through overgrazing and erosion resulting in growing water shortages and desertification. Vast tracks of rainforests are being cut down for conversion to grazing including, according to the FAO report, some 70 percent of the former forests of the Amazon Basin.

Furthermore, animal wastes, chemicals, fertilizers, pesticides, hormones, phosphorous and nitrogen contamination, all byproducts of livestock production, are contaminating water supplies, causing euthropication of lakes (including Lake Winnipeg in Canada) and loss of biodiversity around the world.[12] Increasingly, animal related illnesses are threatening the general population, from bird flu to food and water poisoning.[13]

While concerning themselves primarily with reducing fossil-fuel use in transportation, environmentalists have only approached the livestock problem with policies of promoting vegetarianism and less meat and dairy consumption. But this is hollow advice. Even if the entire developed world were to reduce its meat and dairy consumption significantly, the rising demand in the developing world would still overwhelm the effect on world consumption. Furthermore, as Cuba has shown, increased use of livestock as draught animals (oxen) was required to replace oil-dependent tractors and as a source of organic fertilizer. What it boils down to is that the basic problem is that the world is overpopulated. Cuba has pioneered an approach to dealing with peak oil and the food supply but not with overpopulation and climate change.

Notes

1. Dale Allen Pfeiffer, "Without Oil, Families Will Go Hungry, Not Just Their SUVs," *CCPA Monitor*, April 2006.
2. James Kunstler, *The Long Emergency: Surviving the Converging Catastrophes of the Twenty-First Century* (New York: Atlantic Monthly Press, 2006), p. 3.
3. "Biodiesel," http://en.wikipedia.org/wiki/Biodiesel, accessed February 5, 2007.
4. Ibid.
5. James Kunstler, James. "Clusterfuck Nation Chronicle," http://www.kunstler.com/mags_diary20.html, accessed February 5, 2007.
6. Ibid.
7. The transformation of Cuban agriculture was given a comprehensive treatment in the documentary by David Suzuki shown on the CBC program *The Nature of Things* in 2006.
8. Micheline Sheehy Skeffington, "Organic Fruit and Vegetable Growing as a National Policy: The Cuban Story," *Energy Bulletin*, February 21, 2006, http://www.energybulletin.net/13067.html.
9. Bill McKibben, "What Will You Be Eating When The Revolution Comes?" *Harper's Magazine*, April 2005, p. 61–9.
10. Megan Quinn, "Interview: A 'Community Solution' for Peak Oil," http://archives.econ.utah.edu/archives/a-list/2005w17/msg00017.html, accessed February 23, 2006.
11. Food and Agricultural Organization, "Livestock a Major Threat to Environment," http://www.fao.org/newsroom/en/news/2006/1000448/index.html, accessed February 5, 2007.
12. Organic pollution of bodies of water causing extreme algae growth and de-oxygenation of the water.
13. The best-known example in Canada was the e-coli contamination of Walkerton, Ontario's water supply in 2000 that killed seven people.

Further Reading

Deffeys, Kenneth. *Hubbert's Peak: The Impending Oil Shortage*. Princeton: University Press, 2001.

Flannery, Tim. *The Weather Makers: How Man Is Changing the Climate and What It Means for Life on Earth*. New York: Atlantic Monthly Press, 2006.

Heinberg, Richard. *The Party's Over: Oil, War and the Fate of Industrial Societies*. Gabriola Island, BC: New Society Publishers, 2003.

Kolbert, Elizabeth. *Field Notes from a Catastrophe: Man, Nature, and Climate Change*. New York:

Bloomsbury, 2006.

Manning, Richard. "The Oil We Eat: Following The Food Chain Back To Iraq." *Harper's Magazine,* February 2004, p. 37–45.

Pearce, Fred. *When the Rivers Run Dry. Water—The Defining Crisis of the Twenty-First Century.* Boston: Beacon Press, 2006

Pfeiffer, Dale Allen. *Eating Fossil Fuels: Oil, Food and the Coming Crisis in Agriculture.* Gabriola Island, BC: New Society Publishers, 2006.

PART 2

CLIMATE CHANGE

Bridging Peak Oil and
Climate Change Activism

by Richard Heinberg

The problems of climate change and peak oil both result from societal dependence on fossil fuels. But just how the impacts of these two problems relate to one another, and how policies to address them should differ or overlap, are questions that have so far not been adequately discussed.

Despite the fact that they are closely related, the two issues are in many respects dissimilar. Climate change has to do with carbon emissions and their effects—including the impacts on human societies from rising sea levels, widespread and prolonged droughts, habitat loss, extreme weather events, and so on. Peak oil, on the other hand, has to do with coming shortfalls in the supply of fuels on which society has become overwhelmingly dependent—leading certainly to higher prices for oil and its many products, and perhaps to massive economic disruption and more oil wars. The first has more directly to do with the environment, the second with human society and its dependencies and vulnerabilities. At the most superficial level, we could say that climate change is an end-of-tailpipe problem, while peak oil is an into-fuel-tank problem.

My own background is primarily as an environmentalist. I teach a college course on human ecology and have been writing about ecological issues for fifteen years or so; at the same time, I find myself identified primarily as a peak oil activist, having written three books about the subject and having given something like 300 lectures on it. To me, head-butting arguments between the two groups as to which problem is more serious constitute a peculiar kind of hell, in that such arguments can only hamper the efforts of both groups in doing what we all agree is essential—averting environmental and human catastrophe. Nevertheless, disagreements and misunderstandings are already emerging for the simple reason that advocates on both issues are competing to persuade the public of the central importance of their cause.

Since such competitive disagreements are ultimately damaging to our broader collective interests, it seems important to devote some effort toward openly discussing the differences and similarities of the issues themselves, as well as the priorities and views of their respective interest groups. My thesis is that both groups are essentially working toward a reduction in society's consumption of fossil fuels, and that cooperative efforts between the two groups could substantially strengthen their arguments and their effectiveness at persuading policymakers.

Differing Perspectives

It is fair to note that some peak-oil analysts seem to be of the opinion that oil depletion constitutes a solution to the dilemma of global greenhouse gas emissions, or that climate change is actually not a problem at all. This appears to be the view primarily of some former oil industry geologists, but is almost certainly not that of the majority of depletion analysts. Nevertheless, it is a notion that understandably causes concern and consternation among climate-change activists.

For their part, many climate-change activists and experts see global warming as potentially having such devastating consequences, not just for humans but for the whole biosphere, that peak oil seems a trivial concern by comparison. They argue that, even if global oil production peaks soon, this will provide no solution whatever to climate change because society will replace oil with coal and other low-grade fossil fuels—which will simply worsen greenhouse gas emissions. Moreover, since the remedies for carbon emissions that climate activists propose will inevitably lead to increased energy efficiency and a reduction in oil consumption, they often feel such efforts constitute an adequate answer to the peak oil problem.

Most oil depletionists (excepting the small group discussed above) appear to hold the opinion that climate change is indeed a legitimate concern; however, since the economic impact of peak oil looms in the immediate future, the economic and geopolitical chaos that may be triggered by declining global fuel supplies pose the more timely threat. Some have argued that if peak oil results in near-term economic collapse and wars over dwindling energy resources, these events will seriously or terminally undermine the ability of national leaders to undertake the cooperative, long-range planning necessary to reduce carbon emissions.

For many climate change activists, theirs is primarily a moral issue having to do with the fate of future generations and other species. Their message implies an appeal to self-preservation, but since they cannot prove that the most horrific climate consequences being predicted (the drowning of coastal cities by rising seas, rapidly expanding deserts, collapsing agricultural production) will occur within the next decade or two, the motive of self-preservation is often downplayed. This emphasis on the moral dimension of climate activism is clear in Al Gore's documentary film, *An Inconvenient Truth*.

It is probably safe to say that most peak oil activists are motivated more by their immediate concerns for preservation of self, family, and community. They see the peak of global oil production as happening soon and the effects accumulating quickly. This concern for self-preservation is prominent in the quasi-survivalist tone of several peak oil websites.

Perhaps because climate-change activists see that a dramatic reduction in emissions must be undertaken voluntarily and proactively, and that the depletion of fossil fuels will not occur quickly enough to deter catastrophic emissions levels, they tend to accept generous estimates of remaining fossil fuels as a way of dramatizing the need for action. They see the argument that depletion will take care of the carbon emissions problem as a threat, because it could lead to apathy. They argue that there are enough fossil fuels left on the planet to trigger a climatic doomsday; and, to underscore the argument, climate-change activists often quote robust estimates of remaining oil reserves and amounts awaiting discovery issued by agencies such as the United States Energy Information Administration (EIA) and by companies like ExxonMobil and Cambridge Energy Research Associates (CERA)—most of whose forecasts seem unrealistically optimistic compared to the majority of expert forecasts.

Peak oil activists adhere to more pessimistic resource estimates and production forecasts, and it is tempting to think that this is partly because doing so makes their case appear stronger. However, the track record of prediction by the optimists is not good; just one example: In their IEO 2003 report, the EIA predicted that the country of Oman was "expected to increase output gradually over the first half of this decade" with "only a gradual production decline after 2005." In fact, Oman's production had already peaked in 2000, three years before the forecast was published. The pattern of unrealistic optimism on the part of the official forecasting agencies continues with regard to other countries. So it might be unrealistic for the climate activists to give credence to such forecasts, official though they may be, or even to assume that the truth lies somewhere equidistant between the extreme resource estimates of the so-called optimists and pessimists.

Differing Recommendations

These differences in perspective lead to somewhat diverging policy recommendations.

For climate-change analysts and activists, emissions are the essence of the problem, and so anything that will reduce emissions is viewed as a solution. If societies shift from using a high-carbon fossil fuel (coal) to a fossil fuel with lower carbon content (natural gas), this an obvious benefit in terms of climate risk—and it is potentially an easy sell to politicians and the general public, because it merely requires a change of fuel, not a sacrifice of convenience or comfort on the part of the general public. And so, again, climate analysts tend to accept at face value official high reserves estimates and production forecasts—in this case, for natural gas.

However, as with oil, production forecasts by the official agencies for natural gas supply have tended to be overly robust. For example, in the US the EIA issued no warning whatever of future domestic natural gas problems prior to the supply shortfalls that became painfully apparent after 2000, as prices more than quadrupled. Nevertheless a few industry insiders had noted disturbing signs: companies were drilling at an accelerating pace in order to maintain production rates, and newer fields (which tended to be smaller) were depleting ever more quickly. By 2003 the US Energy Secretary was proclaiming a natural gas crisis. In the following three years, warm weather (perhaps due to climate change) and demand destruction (from the off-shoring of many industrial users of natural gas due to high domestic prices) led to a partial relaxing of prices and general complacency. However, US domestic production appears set to decline further, and likely at a rapid pace.

For depletion analysts and activists, societal dependence on vanishing, non-renewable energy resources is the essence of the greatest dilemma that our society currently faces. We have created a complex, global economic infrastructure built to run on fuels that will start to become scarce and expensive very soon. From this perspective, natural gas is not a solution but an enormous problem: even if the global peak in gas production is ten to twenty years away, regional shortages are already appearing and will continue to intensify. This means enormous risks for home heating, for the chemicals and plastics industries and for electrical power generation. Natural gas is and will always be a fuel that is, for the most part, regionally traded (as opposed to liquid fuels, which are more easily shipped). Thus for many nations critical to the world economy—the US, Britain and most of continental Europe—gas cannot serve as a "transition fuel."

Coal presents another controversial topic for both depletion and emissions analysts. Most members of both groups feel a keen need to articulate some politically palatable transition strategy so as to gain the ears of policy makers. If coal were entirely ruled out of the discussion, such a strategy would become more difficult to cobble together. However, the two groups tend to think of very different future roles for coal.

Some emissions activists and analysts look to so-called "clean coal" as a partial solution to the problem of climate change. Clean coal practices include gasifying coal underground, in situ, and then separating the resulting greenhouse gases (carbon dioxide and carbon monoxide) and then burying these in ocean sediments or old oilfields or coalmines. This theoretically allows society to gain an energy benefit while reducing additions to atmospheric greenhouse gases.

Many depletion analysts are skeptical of such carbon-capture schemes, believing that, when the world is mired in a supply-driven energy crisis, few nations will be adequately motivated to pay the extra cost (in both financial and energy terms) to separate, handle, and store the carbon from coal; instead they will simply burn whatever is available in order to keep their economies from crashing.

Some depletionists see the world's enormous coal reserves as a partial supply-side answer to peak oil. Using a time-proven process, it is possible to gasify coal and then use the resulting gases to synthesize a high-quality diesel fuel. The South African company Sasol, which has updated the process, is currently under contract to provide several new coal-to-liquids (CTL) plants to China and has announced a plant in Montana.

CTL is not attractive to emissions analysts, however. While some carbon could be captured during the gasification stage (at a modest energy cost), burning the final liquid fuel would release as much carbon into the atmosphere as would burning conventional petroleum diesel.

A few depletion analysts tend to take a skeptical view of future coal supplies. According to most widely quoted estimates, the world has at least two hundred years' worth of coal—at current rates of usage. However, factoring in dramatic increases in usage (to substitute for declining oil and gas supplies), while also taking account of the Hubbert peak phenomenon and the fact that coal resources are of varying quality and accessibility—as the Energy Watch Group in Germany has done—leads to the surprising conclusion that a global peak in coal production could arrive only a decade from now.[1] That raises the question: does it make sense to place great hope in largely untested and expensive carbon sequestration technologies if the new infrastructure needed will be obsolete in just a couple of decades? Imagine the world investing trillions of dollars and working mightily for the next twenty years to build hundreds of "clean" coal (and/or CTL) plants, with the world's electrical grids and transportation systems now becoming overwhelmingly dependent on these technologies, only to see global coal supplies rapidly dwindle. Would the world then have the capital to engage in another strenuous and costly energy transition? And what would be the next energy source?

Other low-grade fossil fuels, such as tar sands, oil shale and heavy oil are also problematic from both the depletion and emissions perspectives. Some depletion analysts recommend full-speed development of these resources. However, the energetic extraction costs for these are usually quite high compared to the energy payoff from the resource extracted. Their already-low energy profit ratio (also known as the energy returned on energy invested, or EROEI) would be compromised still further by efforts to capture and sequester carbon, since, as with coal, these low-grade fuels have a high carbon content as compared to natural gas or conventional oil. Currently, natural gas is used in the processing of tar sands and heavy oil; from an emissions point of view, this is rather like turning gold into lead. Many depletionists point out that, while the total resource base for these substances is enormous, the rate of extraction for each is likely to remain limited by physical factors (such as the availability of natural gas and fresh water needed for processing), so that synthetic liquid fuels from such substances may not help much in dealing with the problem of oil depletion in any case.

Supply Side, Demand Side

By now a disturbing trend becomes clear: the two problems of climate change and peak oil together are worse than either by itself. Strategies that might help to keep lights burning and trucks moving while reducing emissions are questionable from a depletionist point of view, while most strategies to keep the economy energized as oil and gas disappear imply increasing greenhouse gas emissions. As we will see, the closer we look, the worse it gets.

As noted above, both groups need to design a survivable energy transition strategy in order to "sell" their message to policy makers. Carbon emissions come from burning depleting fossil fuels, the primary energy sources for modern societies. Thus both problems boil down to energy problems—and energy is essential to the maintenance of agriculture, transportation, communication and just about everything else that makes up the modern global economy.

With regard to both problems there are only two kinds of solutions: substitution strategies (finding replacement energy sources) and conservation strategies (using energy more efficiently or just doing without). The former are politically preferable, as they do not require behavioral change or sacrifice, though they tend to require more planning and investment. The least palatable option, from a political standpoint, is also the quickest and cheapest—doing without (curtailing current usage). We have gotten used to using enormous amounts of energy, at rates unprecedented in history. If we had to use much less, could we maintain the levels of comfort and economic growth that we have become accustomed to? Could we even keep the lights on?

Several questions become critical: How much of a change in energy supply will be imposed by the peaking of production of oil and natural gas? How much will be required in order to minimize climate change? And how much of that supply shortfall can be made up for with substitution and how much with efficiency, before we have to resort to curtailment?

Climate analysts agree the world needs to reduce emissions considerably. In 1996 the European Environment Council said that the global average surface temperature increase should be held to a maximum of 2°C above pre-industrial levels, and that to accomplish this the atmospheric concentration of carbon dioxide will have to be stabilized at 550 parts per million (the current concentration is 380 ppm, though the addition of other greenhouse gases raises the figure to the equivalent of 440 to 450 ppm of carbon dioxide). But recent studies have tended to suggest that, in order to achieve the 2°C cap, much lower carbon dioxide levels will be needed. One study by researchers at the Potsdam Institute for Climate Impact in Germany concluded that—again, to keep the temperature from increasing more than 2°C—the atmospheric concentration target should be 440 ppm of carbon dioxide equivalents. This implies that the atmospheric concentration of greenhouse gases will need to be stabilized at current levels. But, to make the challenge even more difficult, it turns out that the biosphere's ability to absorb carbon is being reduced by human activity, and this must be factored into the equation; by 2030, this carbon-absorbing ability will have been reduced from the current 4 billion tons per year to 2.7 billion. Thus if an equilibrium level of atmospheric carbon is to be maintained through 2030, emissions will have to be reduced from the current annual level of 7 billion tons to 2.7 billion tons, a reduction of 60 percent. It is hard to imagine how, if that translated to a 60 percent reduction in energy consumption, it could mean anything but economic ruin for the world.

Depletion analysts look to about a 2 percent per year decline in oil extraction following the peak of global oil production, with the rate increasing somewhat as time goes on. Regional natural gas decline rates will be much steeper. The dates for global production peaks for both fuels, and for coal, are of course still a matter for some speculation; however, it is reasonable to estimate that we might see somewhere between a 25 and 40 percent decline in energy available to the world's growing population over the next quarter-century as a result of depletion.

Everyone would be happy if it were possible simply to substitute renewable sources of energy for oil, coal, and gas, and both depletion activists and climate activists support the expansion of most renewable energy technologies, including solar and wind. But there are realistic limits to the scale at which renewables can be deployed, and to the speed with which this can be accomplished.

Not all depletion or emissions activists support the large-scale development of biofuels (ethanol, butanol and biodiesel), which are the only realistic renewable replacements for liquid transport fuels, because of the low EROEI entailed in making these fuels, and because these substitutes imply worrisome tradeoffs with food production.

Some depletionists and some climate analysts recommend expanding nuclear power, arguing that technological advances could make it a safe and affordable alternative. Others argue against it, noting that high-grade ores will be depleted in sixty years, and that the entire nuclear cycle of mining, refining, enrichment, plant construction and so on (excluding fission itself) is carbon intensive. One analysis suggests that, from the mid 2020s, the task of clearing up all past and future nuclear wastes will require more energy than the industry can generate from the remaining ore.[2]

Where does this leave us? Let's assume that the more pessimistic critical analyses of both groups are correct. That is, let's say that a 60 percent reduction in emissions is needed within twenty-five years, that natural gas will not be available in sufficient quantities to serve as a transition fuel, that clean coal will not help much, that low-grade fossil fuels will not make up for shortfalls in oil production, that CTL production will (or should) remain marginal, that renewables will not come on line in sufficient quantity or soon enough, that nuclear power won't come to the rescue—and that modest contributions from these sources added together will not come close to making up for shortfalls from oil and gas depletion or from the voluntary phasing out of carbon fuels.

If this turns out to be the case, we may face a staggering need for energy efficiency and curtailment. Neither group wants this as its political platform.

Common Ground

As we have seen, there are understandable reasons for some climate activists to ignore the arguments and priorities of depletionists, and vice versa. Dealing with only one of the two problems is much easier than confronting both. But our goal must be to deal with reality (rather than merely our preferred image of reality), and reality is complicated. Our world faces the interacting impacts not only of peak oil and climate change, but also of water scarcity, overpopulation, over-fishing, chemical pollution and war (among others). In the end, there are too many of us using too much too fast, while competing for dwindling resources.

What would it take to solve all of these problems at once? A good start would be to require a global across-the-board 5 percent per year reduction in fossil-fuel consumption and the provision of substantial financial and technical aid by industrialized nations to less-industrialized nations in creating a renewable energy infrastructure. But to the patient (the primary fossil-fuel users) this medicine might seem worse than the disease. A grand plan like this has almost no chance of gaining political backing.

Realistically, we are left with the customary policy tools aimed to ameliorate the world's ills piecemeal: emissions and depletion protocols, tradable quotas, emissions rights, import and export quotas, carbon taxes and cap-and-trade mechanisms.

So for practical reasons it is probably inevitable that emissions and depletion activists will continue to pursue their separate policy goals. But it makes sense for the two groups to be informed by one another, and to cooperate wherever possible.

It is fairly obvious why such cooperation would benefit the depletionists: climate change is already a subject of considerable international concern and action, whereas peak oil is still a relatively new topic of discussion.

But how would such cooperation aid emissions activists?

In a word: motivation. As discussed earlier, emissions activists appeal to an ethical impulse to avert future harm to the environment and human society, while the peak oil issue appeals to a more immediate concern for self-preservation. In extreme circumstances, the latter is unquestionably the stronger motive. Strong motivation will certainly be required in order for the people of the world to undertake the enormous personal and social sacrifices required in order to quickly and dramatically reduce their fossil-fuel dependency. Sustainability and equity are issues that are hard enough to campaign on in times of prosperity; when families and nations are struggling to maintain themselves due to fuel shortages and soaring prices, only massive education and persuasion campaigns could possibly summon the needed support.

Taken together, climate change and peak oil make a nearly air-tight argument. We should reduce our dependency on fossil fuels for the sake of future generations and the rest of the biosphere; but even if we choose not to do so because of the costs involved, those fuels will soon become more scarce and expensive anyway, so complacency is simply not an option.

What would cooperation between the two groups look like? It would help, first of all, for activists on one issue to spend more time studying the literature of the other, and for both groups to arrange meetings and conferences where the intersections of the two issues can be further explored.

Both groups could work together more explicitly to promote proactive, policy-driven reductions in fossil fuel consumption.

Climate activists could start using depletion arguments and data in tandem with their ongoing discussions of ice cores and melting glaciers, but to do so they would need to stop taking unrealistically robust resource estimates at face value.

For their part, depletionists—if they are to take advantage of increased collaboration with emissions activists—must better familiarize themselves with climate science, so that their peak oil mitigation proposals are ones that lead to a reduction rather than an increase of carbon emissions into the atmosphere.

Perhaps, for both groups, with a stronger potential for motivating the public will come the courage to tell a truth that few policy makers want to hear: energy efficiency and curtailment will almost certainly have to be the world's dominant responses to both issues.

Notes

1. Energy Watch Group, "Coal: Resources and Future Production," www.energywatchgroup.org/files/Coalreport.pdf.
2. David Fleming, *The Lean Guide to Nuclear Energy: A Life-Cycle in Trouble*. London: The *Lean economy Connection*, 2006. See also: Energy Watch Group, "Uranium Resources and Nuclear Energy," http://www.energiekrise.de/news/docs/specials2006/REO-Uranium_5-12-2006.pdf.

Further Reading

Peak Oil information and perspectives:

Kenneth Deffeyes, *Beyond Oil: The View from Hubbert's Peak* (Hill and Wang, 2005)
Richard Heinberg, *The Party's Over: Oil, War and the Fate of Industrial Societies* (New Society, 2005)
_____ , *The Oil Depletion Protocol: A Plan to Avert Oil Wars, Terrorism and Economic Collapse* (New Society, 2006)
http://www.theoildrum.com
http://www.energybulletin.net
http://www.globalpublicmedia.com

Peak Oil activism:

http://www.postcarbon.org
http://www.oildepletionprotocol.org
http://www.communitysolution.org

Climate Change information and perspectives:

Ross Gelbspan, *Boiling Point* (Basic Books, 2005)
Al Gore, *An Inconvenient Truth: The Planetary Emergency of Global Warming and What We Can Do About It* (Rodale Books, 2006)
George Monbiot, *Heat: How to Stop the Planet from Burning* (South End Press, 2007)
http://www.realclimate.org
http://climatewire.org/
http://illconsidered.blogspot.com/

Climate activism:

http://www.cleartheair.org
http://www.climateprogress.org
http://reasic.com

Getting Personal about Climate Change

by Metta Spencer

If we are serious about limiting global climate change, we're going to face some disagreeable social dilemmas. Every social movement is prodded along by "moral entrepreneurs"—gadflies who challenge previously accepted ways of living. In this case, that means promoting new, draconian measures to reduce the use of fossil fuels. But personal confrontations are not nice, so we avoid them. Thirty years ago I didn't object when others smoked around me. Only non-smokers braver than I took responsibility for establishing a new social norm. Nor did I ever refuse to serve coffee to male colleagues. I let other feminists speak up against gender discrimination. So now it's my turn—and yours—to be militant. Some of us have to become moral entrepreneurs, changing our personal habits of consumption and—especially—assertively urging changes in others and in our political and economic institutions.

The International Panel on Climate Change (IPCC), with its 2500 scientists, has been releasing its fourth report in several segments in 2007,[1] and the projections are, as expected, grave. Our challenge is unique in human history. We confront a crisis that has come upon us gradually, imperceptibly, but suddenly has become urgent and life threatening.

The Crisis

The planet's climate is changing, largely because of human activities, especially our use of fossil fuels, which produce greenhouse gases (GHG): methane, nitrous oxide, fluorinated and chlorinated gases, ground-level ozone and, most notably, carbon dioxide, which trap heat in the atmosphere.

Throughout most of human history, atmospheric carbon dioxide was about 280 parts per million by volume (ppmv), but now it approaches 400 ppmv. The other GHGs have also increased dramatically.[2] According to some scientists, these are the highest levels in twenty million years.[3]

Between 1750 (when the Industrial Revolution began) and the present, GHG emissions have grown, with an increase of fully 70 percent between 1970 and 2004.[4] Accordingly, the earth's temperature has increased by about 0.6°C. Our challenge is to limit the earth's additional increase to 1.4°C above the present level, or 2°C above pre-industrial levels.[5] Unfortunately, it is virtually certain that the twenty-first century will see additional warming over most land areas. We can expect increasing extremes of weather, including droughts, cyclones, high winds and high sea levels. For the next two decades, a warming of 0.2°C per decade is projected.[6] Even if GHGs could be kept at present levels, a continuing warming of 0.1°C per decade would occur because the effects on the oceans take a long period and some gases remain active for as long as a century.

As carbon dioxide concentrations increase, the oceans are becoming increasingly acidified. Widespread melting of glaciers, ice caps and the ice sheets of Greenland and Antarctica contributed to the faster rising sea level between 1993 to 2003. There is a possibility that 90 percent of the top ten feet of permafrost will thaw during this century.[7]

Other factors of both human and natural origin also affect the climate. For example, aerosols (e.g., from volcanic eruptions) had a cooling effect during the mid twentieth century, mitigating the increasingly warm trend for a while, despite also depleting the stratospheric ozone. Among the other warming factors is animal agriculture, the source of 15 percent of the world's greenhouse gas emissions—especially nitrous oxide and methane.[8] Deforestation also exacerbates the warming because trees, along with oceans, serve as "sinks" that sequester carbon dioxide.[9]

There are many complex interactions among the determinants of climate—including feedback loops that may exacerbate or reverse trends. For example, snow and ice reflect sunlight, thereby helping to keep the planet cool (the "albedo effect"). The melting of icecaps exposes dark land and water to sunshine, thus absorbing rather than reflecting solar heat. Another example: immense fields of a GHG, frozen methane, lie beneath the oceans and polar soils. When permafrost melts, the methane below it can also melt and enter the atmosphere, compounding the warming effect. It is about twenty times as potent a greenhouse gas as carbon dioxide.[10]

The IPCC report has projected ranges of warming during the twenty-first century for six alternative scenarios of GHG emissions. The worst scenario would show a temperature rise of 4°C and a sea level increase between 0.26 and 0.59 meters. The most favorable model would give a temperature change of 1.8°C and a sea level increase between 0.18 and 0.38 meters. Even in such a "favorable" outcome, many millions of people will lose their homes to floods.[11]

No person and no society will escape the effects of climate change. Probably it is already too late to save the world's coral reefs, which are the habitat of about a million species.[12] Low-lying coastal lands will all be especially vulnerable. For example, the homes of some 60 million human beings living in the vicinity of Calcutta may be under water if the temperature increases by 2°C or more, the tipping point into intolerable and irreversible changes.

The World Health Organization estimates that 150,000 persons are already dying each year of diseases that can be attributed to climate change.[13]

The Arctic is particularly affected, so Canada is already experiencing some of the early effects of global warming. For example, caribou herds and polar bears are declining. Nevertheless, Canada continues to emit more GHGs per capita than any other nation. All other industrial societies, including the United States, have been trying to reduce GHGs—but, as of early 2007, not Canada. Canada originates 5.5 tonnes of carbon emissions per person per year,[14] whereas, according to George Monbiot's convincing calculations, our biosphere can be sustainable only if the average person on earth limits emissions to about 0.33 tonnes per year.[15]

A major explanation for Canada's bad record is the mining of tar sands in Alberta, which yield a substitute for petroleum. Regrettably, the oil product itself, as well as the fuel used to melt the tar and separate it from the sand, emit vast quantities of carbon dioxide—five times as much as the production of conventional oil—and require huge quantities of water from underground aquifers.[16] Australia and the United States, as indeed all other democratic societies besides Canada, have decided against producing this dirty oil. Nevertheless, Canada cannot curtail tar sands development since, under the terms of the North American Free Trade Agreement, Canada is obliged to export the same proportion of energy to the United States as it has over the preceding three years.

Meanwhile, with the publication of the IPCC's fourth report, the crucial scientific questions have been answered. Very few people still dispute the IPCC conclusions. Lawrence Solomon wrote a series profiling ten scientists who had questioned some technical aspect of the third IPCC report.[17] None of them, however, denied the reality of global warming or that it results largely from human activity. Only one big question still has to be decided, namely: *What shall we do about it?*

Looming Political Controversies

Like the George W. Bush administration in the United States, Canada's Conservative government has decided against trying to fulfill the GHG objectives of the Kyoto Protocol, claiming that it would cause an economic collapse comparable to that experienced by Russia after the fall of Communism. However, a different opinion was expressed in October 2006 by the former chief economist and senior vice-president of the World Bank, Sir Nicholas Stern, who predicted that climate change could cut every country's economy by one-fifth and that only drastic interventions can prevent that outcome.[18] He argues that delay will cost far more than timely action. Still, financial factors were considered so compelling that not until 2007 did either the

United States or Canadian governments begin responding to the growing political pressure for change.

For the George W. Bush administration, this lag reflected the priority assigned to war. According to Nobel laureate economist Joseph Stiglitz, the accrual cost of the Iraq War (i.e., the total long-term committed cost, not just the amount authorized by Congress) is $2.3 trillion.[19] For this amount of money, the country could have installed wind turbines providing about 10 kW for every US citizen—enough to generate all the electricity now being used in the United States, with extra left over to run electric cars or to produce hydrogen as a transportation fuel.

To limit the earth's warming, almost all nations have ratified the Kyoto Protocol, an amendment to the United Nations Framework Convention on Climate Change, which assigns mandatory targets for the reduction of greenhouse-gas emissions to signatory nations. A 1997 treaty, Kyoto requires all industrialized countries to mitigate climate change by reducing emissions of GHGs and protecting sinks and reservoirs—the forests and bodies of water that store carbon. Yet the targets that countries have are not all equally rigorous. France aims to reduce emissions by 75 to 80 percent by the year 2050. California has set its own reduction target for that year: 80 percent. Canada's National Round Table on the Environment and the Economy proposed a reduction of GHG emissions by 60 percent, though environmental organizations insist that this target is too low. To be safe, 90 percent is a better goal.[20]

Competing Solutions

Fortunately, renewable energy is abundant. What is scarce is the time required for harnessing it in sufficient quantities. Decades will be required, and we face painful political dilemmas. For example, although few nuclear power plants have been constructed lately because of the dangers they pose, there are now calls for additional reactors since they do not directly emit GHGs. Unfortunately, the mining and processing of fuel for nuclear plants *does* emit a lot of carbon dioxide. Alternatively, gas could be exploited more to replace coal or oil, since gas does not contribute as seriously to global warming. It may buy us enough time to develop genuinely renewable sources of energy.

A number of countries are already participating in a type of rationing called "cap-and-trade." GHG emission limits or "caps" are set for all industries, with specific credits allocated to each organization. If a firm fails to use up its allocated emissions, it can sell the surplus back to the system, so that a different company (perhaps even in a different country) can buy those rights. Thus the buyer is, in effect, fined for polluting, whereas the seller is rewarded financially for reducing emissions. Over time, the system's administrators will gradually reduce the caps. Countries signatory to the Kyoto Protocol have been assigned caps. In 2003, corporations began trading GHG emissions voluntarily on the Chicago Climate Exchange, and European Union nations have their own trading system underway. This system has been used successfully in controlling other types of pollution, such as reducing acid rain. However, it has not yet been effective in reducing GHGs. Some economists, including Alan Greenspan, former chairman of the US Federal Reserve, believe that it cannot. According to Greenspan, to have dramatic effects on industry as a whole, a cap would have to be set so low that "companies could not buy their way out of having to make production changes."[21]

In the years ahead much deeper cuts in emissions must be made, not just by businesses, but also by individuals and households. Without publicly committing to it yet, the British government has quietly been exploring a plan for rationing emission credits.[22] Individuals would receive a card each year. Thereafter, when buying a fossil fuel by, say, purchasing an airplane ticket, filling up a car's gas tank or ordering heating oil for a home the corresponding number of credits would be subtracted. This system would presumably increase socio-economic equality somewhat, for the poor are less likely to use up their emission credits than the well-to-do, who will pay dearly for the privilege of living extravagantly.

Still, it seems unlikely that most societies will adopt rationing schemes. An alternative is simply to let the market price of fuels determine the distribution of gasoline, electricity, heating oil and the finished goods and services that depend on these fuels. Agriculture, for example, now uses a lot of oil in producing food and even more for transporting it to stores and hence to our homes. If oil prices increase dramatically—say as a result of heavy carbon taxes—those other commodity prices will increase correspondingly and consumers will begin rationing themselves by revising their own household budgets. In principle, rationing is unnecessary if the market is allowed to function without subsidies and if prices are kept high. However, such a free-market system worsens the hardships of low-income families.

Each society will choose among these differing approaches through its own political processes, depending on its values. For example, military activities account for a sizeable portion of GHG emissions. An F-16 fighter jet uses twice as much fuel in one hour as the average motorist consumes in a year. Emissions from military operations account for 6 to 10 percent of global air pollution. In 1988, Pentagon activities produced 46 million tons of carbon—3.5 percent of the US total. When the need for cutbacks become urgent, citizens whose values are peace-oriented may force their governments to reduce military activities for the sake of a habitable climate.[23]

In Canada, an especially contentious issue involves whether to permit the continued development of the Alberta tar sands. No issue is more likely than this to pit environmentalists against economic interests like the companies heavily invested in the tar sands. As well, the labour shortage in Alberta has enabled unemployed workers throughout Canada to move to that province and find jobs. There will be great resistance, therefore, to any proposal to shut down the tar-sands operation—a policy that has must be adopted if Canadians are to stay within the necessary limits.

Governments will be deeply involved in numerous other programs that will determine the success or failure of our global future. For example, there are already debates about whether new electrical power plants should use nuclear power, say, or wind turbines, or old-fashioned coal with its carbon emissions sequestered in abandoned mines or at the bottom of oceans. Some governments are now offering special loans for families and businesses to install solar panels and other renewable sources of energy, as well as insulation and other retrofitting schemes. It is governments that create new transportation systems, as well as zoning ordinances and building codes that determine the shape of the towns we will inhabit. About one-third of energy use goes into heating and cooling buildings. Therefore, setting higher standards for new buildings and incentives for upgrading older buildings can have a huge impact.

Technological Prospects

Our way of life is about to change abruptly in response to radical technological developments. We require vast breakthroughs in efficiency, which are forthcoming as new models of appliances, vehicles and housing appear, reducing fuel requirements. But besides such incremental means of conservation, there must also be radical and surprising new inventions that generate large quantities of renewable sources of energy.[24]

There is one major obstacle: At present, most renewable energy is of low intensity or low quality. For example, an immense amount of solar radiation strikes the earth's surface—but it is dispersed over such a wide area that it is hard to collect in sufficient quantities. Likewise, wind and sunshine are intermittent, not consistently available, and batteries are not yet able to store much energy for later use. Large amounts of energy must be invested in order to produce and deliver adequate renewable energy. Indeed, for some products, such as ethanol made from corn, the output energy only slightly exceeds the amount of the input energy. Nevertheless, ethanol production is booming in the United States and is expected to become profitable as soon as cellulosic waste products, such as straw and switch grass can become the main feedstock.

Fossil fuels are richer than most renewables as sources of energy—or at least they were until depletion gradually made them harder to obtain. Even now, the net energy of coal is high—its energy output is twenty-five times greater than the energy input that produces it. Therefore, to shift from coal to most of the existing renewable fuels would be unprofitable. But coal is the dirtiest source of energy; its GHG emissions are appalling, even when it is converted to liquid fuel. The most urgent necessity is to develop new alternative technologies that offer vastly superior net energy than current ones.

Already, the net energy of wind turbines is superior to most wells. New conventional wells have a high net energy, but they are few and far between and small. Most new wells are deep ocean, or arctic or tar sands—all with much lower net energy than oil wells. However, wind's intermittency is a drawback. One solution now being developed is to tether big kite-like turbines high in the jet streams, where wind is always strong.[25] These kites can generate enormous energy. Their only significant disadvantage is that the tether is a perfect lightning rod, so they have to be pulled down whenever a storm approaches.

Still other radically new technologies are being developed. One involves the sequestration of carbon dioxide by single-cell organisms: algae. There are many different species of algae, including some composed largely of oil that thrive on carbon dioxide and sunlight.[26] A Boston company's method involves running the emissions from a local power plant flue through clear plastic tubes full of water and fast-growing algae. Every day, some of the algae are harvested and squeezed, yielding bio-diesel fuel plus some dry green flakes that have other uses. While moving through the plastic tubes, almost half of the carbon dioxide and 85 percent of the nitrous oxide is removed from the flue's emissions. Thus the system triumphs in two ways: both by removing GHGs and by generating a substitute fuel.

Another promising technology uses "concentrating solar power" (CSP), a process with the potential to generate enough electricity for the entire planet from sunlight.[27] Its principle is entirely different from the more familiar photovoltaic solar cell. It is a mature technology, for one such installation has been operating in the California desert for twenty years. Now a huge CSP electric power plant is being built in the Sahara. A tower will stand in an open field, surrounded by hundreds of mirrors that track the sun and direct its rays toward the tower where a fluid is flowing and heating up to 1000°C. This heat will be transferred either to a power generator or into molten salt, where it can be stored for a few hours at night. This is a highly efficient source of electricity, which can be sent to Europe as high-voltage direct current.

An altogether different technological solution to the problem of global warming involves the capture of ambient carbon dioxide from the planet's atmosphere. Klaus S. Lackner, a Columbia University physicist, has developed a device in which sorbents capture carbon dioxide molecules from free-flowing air for sequestration. Since the carbon dioxide in the atmosphere is evenly distributed around the globe, it is possible to set up such devices at any convenient location, not just at smokestacks or exhaust pipes where GHGs are emitted. Global Research Technologies has tested a prototype of the system and will develop it for commercial production. According to Physorg.com,

> A device with an opening of one square meter can extract about 10 tons of carbon dioxide from the atmosphere each year. If a single device were to measure 10 meters by 10 meters it could extract 1,000 tons each year. On this scale, one million devices would be required to remove one billion tons of carbon dioxide from the atmosphere. According to the U.K. Treasury's Stern Review on climate change, the world will need to reduce carbon emissions by 11 billion tons by 2025…. Air capture devices are small and require much less land area than the wind mills that would be needed to offset an equal amount of CO_2 emission.[28]

Personal Decisions

We cannot wait for these technological and political developments. As individuals we need to begin changing our lifestyles. George Monbiot's book, *Heat: How to Stop the Planet from Burning*, suggest how to priorize the various changes available today. He suggests, for example, that we consume locally grown food rather than squander energy on transport. Fortunately, it is possible to grow lots of produce in cold climates without using a heated greenhouse.

Monbiot also suggests that we create an inter-urban bus system with routes through the city, picking people up in several sites so passengers don't have to burn energy going to centralized terminals. Likewise, he would do away with most grocery stores. Today, food is transported from warehouses to local shops, which must be heated and well-lighted. Then we use our cars to go grocery shopping. Instead, Monbiot favors the system that I've been using for years—shopping by Internet. Each order is packaged up in the warehouse and delivered directly to my kitchen, eliminating the retail store and my trips to it.

Only the challenge of "sustainable flying" leaves Monbiot at a loss. Aviation has been growing by 5 percent a year since 1997, and there are no technological breakthroughs on the drawing board to reduce emissions. Propeller-driven planes are better than jets, but there are few such airliners of that type. By Monbiot's calculations, we will have to reduce air travel by 96 percent. Go by train, bus or ship and, if you visit distant loved ones, expect to go rarely and to stay there a long time. As a fervent moral entrepreneur, he writes bluntly that "long-distance travel, high speed and the curtailment of climate change are not compatible. If you fly, you destroy other people's lives."

Such appeals to our ethical idealism are necessary. We all must become moral entrepreneurs, for habits will not change unless they are forthrightly challenged. Still, this does not have to demoralize anyone. There is another way to look at it. The basis for living a full, rich life is not ease or comfort or affluence, but meaning. We all want to contribute to something larger than ourselves, but ways of doing so are often obscure. Climate change is not obscure. It's real and it's happening now. It presents an opportunity for every human being on the planet to become a hero.

Notes

1. Intergovernmental Panel on Climate Change, *Working Group I Fourth Assessment Report: "The Physical Science Basis"; Working Group II Fourth Assessment Report, Climate Change 2007: "Impacts, Adaptation and Vulnerability"; Working Group III, Fourth Assessment Report, Climate Change 2007: "Mitigation of Climate Change."*

2. M. Meinhausen, "What Does a 2 Degree Target Mean for Greenhouse Gas Concentrations?" in Hans Joachim Schellnhuber, Wolfgang Cramer and Nebojsa Nakicenovic, eds., *Avoiding Dangerous Climate Change* (Cambridge: Cambridge University Press, 2006), p. 392.

3. Paul N. Pearson and Martin R Palmer, "Atmospheric carbon dioxide concentrations over the past 60 million years," *Nature,* August 17, 2000. Also, ice cores from Antarctica contain levels of carbon dioxide and methane that are higher than ever before in the past 650,000 years. See Urs Siegenthaler et al., "Stable Carbon Cycle-Climate Relationship During the Late Pleistocene," *Science,* November 25, 2005, pp. 1313–17.

4. IPCC Fourth Assessment Report, *Working Group III,* "Summary for Policymakers," p. 3.

5. George Monbiot, *Heat: How to Stop the Planet from Burning* (Toronto: Doubleday Canada, 2006), p. 15. Monbiot reviews the research justifying this projection.

6. For an overview of projected effects, see ibid., p. 5.

7. National Center for Atmospheric Research, "Most of Arctic's Near-Surface Permafrost May Thaw by 2100," http://www.ucar.edu/news/releases/2005/permafrost.shtml, accessed December 19, 2006.

8. Center for International Earth Science Network, Columbia University, "Methane Emissions from Livestock," http://www.ciesin.columbia.edu/TG/AG/liverear.html.

9. Karl Mallon, Greg Bourne and Richard Modd, *Climate Solutions: WWF's Vision for 2050.* Paper prepared for World Wildlife Fund's Global Energy Task Force, p. 3.

10. Methane accounts for about 19 percent of the GHG emissions into the environment. A catalytic converter could convert it into carbon dioxide with some potential advantages. See "Prof Creates Catalytic Converter to Reduce Greenhouse Gas Emissions," University of Alberta report by Technology, Entrepreneur and Company Development, http://www.uofaweb.ualberta.ca/tecedmonton/news.cfm?story=32655.

11. IPCC *Fourth Assessment Report, Working Group III,* "Summary for Policymakers," pp. 12–13.

12. *Report of Global Marine Species Assessment* (GMSA), a partnership between Conservation International and the World Conservation Union, June 15, 2007, http://www.edie.net/news/news_story.asp?id=13156&channel=0.

13. World Health Organization, *Climate Change,* 2003, http://www.who.int/heli/risks/climate-

change/en/index.html.

14. Sierra Club of Canada, reported at http://www.av-a.org/climate.html.

15. Monbiot, *Heat*, p. 16.

16. Paul Chastko, *Developing Alberta's Oil Sands* (Calgary: University of Calgary Press, 2004), pp. 100–12.

17. Lawrence Solomon, *National Post*, November 28, 2006.

18. Sir Nicholas Stern, *The Economics of Climate Change*, Report to Treasury of UK government, 2006. "Summary of Conclusions," p. vi, http://www.hm-treasury.gov.uk/media/3/2/Summary_of_Conclusions.pdf.

19. Tom Regan, "Report: Iraq War Costs Could Top $2 Trillion," *Christian Science Monitor,* January 10, 2006, http://www.csmonitor.com/2006/0110/dailyUpdate.html.

20. The case for this high figure is argued brilliantly by George Monbiot in his book *Heat*, probably the best practical review of the world's climate-change challenge yet published.

21. *Globe and Mail*, February, 12, 2007, p. B7. For a serious critique of the Kyoto emission trading scheme, as well as the CDMs, see Kevin Smith's contribution in this book and "Carbon Trading: A critical conversation on climate change, privatization and power," *Development Dialogue* 48 (September 2006).

22. Gwynne Dyer, "Getting Radical About Climate Change: The Shape of Things to Come," http://www.gwynnedyer.net/articles/Gwynne%20Dyer%20article_%20%20Getting%20Radical%20About%20Climate%20Change.txt, accessed December 14, 2006.

23. Michael Renner, *State of the World 2000* (Washington, DC: World Watch Institute, 2000).

24. An especially encouraging view of these prospective efficiencies can be found in Ernst von Weizsaker, Amory B. Lovins and L. Hunter Lovins, *Factor Four: Doubling Wealth, Halving Resource Use* (London: Earthscan, 2006). See also Alex Steffen, ed., *World Changing: A User's Guide for the 21st Century* (New York: Abrams, 2006).

25. Keay Davidson, "Scientists Look High in the Sky for Power: Jet Stream Could Fill Global Energy Needs," *San Francisco Chronicle*, May 7, 2007, http://sfgate.com/cgi-bin/article.cgi?file=/c/a/2007/05/07/MNGNEPMD801.DTL.

26. Mae-Wan Ho, "Green Algae for Carbon Capture and Biodiesel, Institute of Science in Society, press release, March 3, 2006, http://www.i-sis.org.uk/GAFCCAB.php.

27. National Renewable Energy Laboratory, "Concentrating Solar Power Research," http://www.nrel.gov/csp/

28. Columbia University's Earth Institute, reported at http://www.physorg.com/news96732819.html.

The Obscenity of Carbon Trading

by Kevin Smith

In 1992, an infamous leaked memo from Lawrence Summers, at the time chief economist of the World Bank, stated that "the economic logic behind dumping a load of toxic waste in the lowest wage country is impeccable, and we should face up to that."[1] The Stern Review on climate change, written by Sir Nicolas Stern, a man who took over Summers' old job from 2000 to 2003, applies a similar sort of free market environmentalism to climate change. Stern argues that the most important factor in making emissions reductions is its cost-effectiveness, advocating mechanisms such as carbon pricing and carbon trading.[2]

While dumping toxic waste might look like a great idea from the perspective of the market, it ignores the glaringly obvious fact that is hugely unfair to the people who live where that toxic waste gets dumped—in the global South. In a similar way, Stern's cost-benefit analysis reduces important debates about the complex issue of climate change down to a discussion about numbers and graphs. In this way, it ignores less quantifiable variables such as human lives, species extinction and widespread social upheaval.

Junk Economics

Cost-benefit analysis can be a useful tool for making choices in relatively simple situations when there are a limited number of straight-forward options to choose from. But in the current neo-liberal economic environment, trading rules succumb to the pressures of corporate lobbying and deregulation. As Tom Burke, visiting professor at Imperial College London, has observed, "The reality is that applying cost-benefit analysis to questions such as [climate change] is junk economics ... It is a vanity of economists to believe that all choices can be boiled down to calculations of monetary value."[3]

Some commentators have applauded the Stern Review for speaking in the economics language that politicians and the business community can understand. But by framing the issue purely in terms of pricing, trade and economic growth, the Stern Review reduces the scope of the response to climate change to market-based solutions.

These solutions take two common forms. The first is the concept that under emissions trading, governments can allocate permits to big industrial polluters to trade a "right to pollute" amongst themselves. The second approach involves the generation of surplus carbon credits. These credits are taken from projects that claim to reduce or avoid emissions in other locations, usually in Southern countries, and then purchased by polluters to make their emissions reduction shortfalls look good.

But both solutions sidestep the most fundamentally effective response to climate change: leaving the fossil fuels in the ground. Even though this is not an easy proposition for our heavily fossil-fuel dependent society, it is precisely what is needed. And therein lies the rub: for what incentive is there to start making these costly, long-term changes when you can simply purchase cheaper, short-term carbon credits?

Forcing the Market

These days, trading rules inevitably succumb to the pressures of corporate lobbying and deregulation, ensuring that governments do not interfere with the smooth running of the market. We have already seen this corrosive influence in the European Union's Emissions Trading Scheme (ETS). Under corporate pressure, governments in the first round massively over-allocated emissions permits to the heaviest polluting industries (Smith). This caused the price of carbon to drop by more than 60 percent, creating even more of a disincentive for industries to lower their source emissions.

What is more, there are now all manner of loopholes and incentives for industry actually to exaggerate their emissions. By doing this, they receive even more permits—and thereby take even less action. For example, market analyst Franck Schuttellar estimated that in the scheme's first year, the UK's most polluting industries collectively earned £940 million ($1946 million CDN) in windfall profits from generous ETS allocations! Given all we know about the link between pollution and climate change, such a massive public concession to dirty industries borders on insanity.

We are continually being asked to believe that the flexibility and efficiency of the market will ensure that carbon is reduced as quickly and as effectively as possible, all the time that experience shows that lack of firm regulation creates, rather than solves, environmental problems.

Pollute and Profit

With the second phase of the EU Emissions Trading Scheme due to start in 2008, Brussels will soon have to admit that the first phase has not met with success. Carbon emissions are not going down, and industries are not switching to clean energy technology. So far, indeed, the scheme's guiding principles seem to have been "polluter profits" rather than "polluter pays." The lack of discernible results suggests that the ETS has been designed for its ideological compatibility with the free market rather than for its effectiveness in achieving urgently needed carbon emissions cuts.

On paper, of course, the "cap and trade" scheme seems foolproof. The amount of permissible carbon pollution is divided up between industrial locations or "installations" across Europe—this is the "cap" part. If any installation pollutes excessively, it must purchase an equivalent amount of permits on the market. Conversely, if an installation pollutes less than the limit, it can sell this shortfall on the market—this is the "trade" part. The cap is supposed to get tighter in successive rounds of the scheme so that the market price of carbon rises, creating the incentive for industries to make low-carbon source modifications rather than having to buy costly permits. The idea is that the market will create the most "cost-effective" reductions possible

But the first phase has been a disaster. One of the main problems of the scheme is that every stage of its design and implementation has been subjected to intensive industry lobbying. As the economist John Kay, writing in the *Financial Times*, has suggested, "when a market is created through political action rather than emerging spontaneously from the needs of buyers and sellers, business will seek to influence market design for commercial advantage."[4]

Under sustained corporate lobbying, almost all EU governments made initially huge over-allocations of permits to industry. In 2005, the first year of trading, for example, the relevant industries across Europe emitted 66 million tonnes less than the allocated cap. This meant that the cap was effectively meaningless: it had not actually forced any net reductions. What is more, a preliminary analysis of the 2006 data shows that 93 per cent of the 10,000 installations covered by the ETS have emitted less than their allotted quota. This represents in 30 million tonnes less than the total EU-wide allocation.[5]

Successful corporate lobbying also meant that permits were initially allocated free of charge to industry. Nonetheless, companies have been ripping consumers off by charging for this non-existent cost anyway. A study by UBS Investment Bank, for example, showed that the first round of the ETS has added 1.3 euro cents to each kilowatt hour of electricity sold. The German minister for the environment estimated that in this way the four biggest power providers in the EU—Eon, RWE, Vattenfall and EnBW—had profited by between $8.5 billion CDN and $11.4 billion CDN from over-allocations and charging consumers the entirely fictional cost of the ETS first phase.[6]

ETS apologists are quick to claim that the second round will ensure that these early "design faults" are ironed out. For starters, governments are now supposed to auction off a percentage of permits to industry rather than simply handing them out for free. Yet in practice, only ten EU members have chosen to go down this route. What is more, four out of the ten are auctioning fewer than 1 percent of their total allocations.[7]

Yet free-allocations to fossil fuel intensive industries continue—in effect, providing a huge subsidy to the heaviest polluters. Dr. Karsten Neuhoff from the Cambridge University Faculty of Economics concluded that "the level of such subsidies under proposed second phase NAP is so high that the construction of coal power stations is more profitable under the ETS with such distorted allocation decisions than in the absence of the ETS."[8]

Advocates of the scheme also argue that the tighter caps imposed in Phase II will cause the price of carbon to increase. This will encourage industries to start implementing cleaner technologies and practices. However, predictions of higher price permits in Phase II are somewhat optimistic in the face of the "linking directive" which means that companies can also acquire credits by investing in Clean Development Mechanism (CDM) projects—that is, offset projects in the global South through the Kyoto Protocol.

This linking directive represents a serious leak in the system that undermines the effectiveness of tightened caps. According to Neuhoff and his colleagues, "some market participants anticipate that the European market could be flooded by these [CDM] allowances to such an extent that the EU allowance price would plummet."[9]

It is not only the availability of such cheap credits that undermine the climate credibility of the ETS; the nature of the CDM projects have come under sustained criticism too. The projects are supposed to bring developmental benefits to local communities, with the market expected to create incentives for investment in low-carbon energy infrastructure in Southern countries. But almost two-thirds of the 1534 CDM projects in the pipeline as of early 2007 did not involve either the generation of clean energy or carbon dioxide emissions.[10]

The largest share of CDM credits (30 percent) has been generated by the destruction of HFC-23. This potent greenhouse gas is created by the manufacture of refrigerant gases. A recent study showed that the value of these credits at current carbon prices was $6.7 billion CDN. Not only was this twice the value of the refrigerant gases themselves, but it was also estimated that the cost of implementing the necessary technology to capture and destroy the HFC-23 was less than $142 million CDN. In other words, something in the region of $6.5 billion CDN was generated in profit for the owners of the plants and the project brokers.[11]

The enormous sum of money generated by these Kyoto-style trading schemes has not gone to the companies and communities who are taking action on clean energy and energy reduction projects, but rather to big, industrial polluters who are then at liberty to reinvest the profits into the expansion of their operations. In the 2006–07 financial year, the owners of SRF, an industrial and textiles company based in India, reported a profit of $123.5 million CDN from the sale of carbon credits derived from the destruction of HFC-23. Ashish Bharat Ram, the managing director, told the *Economic Times* that "strong income from carbon trading strengthened us financially, and now we are expanding into areas related to our core strength of chemical and technical textiles business."[12]

Many of the corporate benefactors of CDM money in Southern countries are the target of sustained local resistance from communities who have to endure the often life-threatening impacts of intensive, industrial pollution. In 2005, about 10,000 people from social movements, community groups and civil society organizations mobilized in Chhattisgarh, India, to protest at the environmental public hearing held for the expansion of Jindal Steel and Power Limited (JSPL) sponge-iron plants in the district.[13] The production of sponge iron (an impure form of the metal) is notoriously dirty, and the companies involved have been accused of land-grabbing, as well as causing intensive air, soil and water pollution.

JSPL runs the largest sponge-iron plant in the world, which is spread over 320 hectares on what used to be the thriving, agricultural village of Patrapali. This plant alone has four separate CDM projects, generating millions of tonnes of supposed carbon reductions that could be imported into the ETS. The inhabitants of three surrounding villages are resisting a proposed 20 billion rupee expansion that would engulf them. The CDM is not only providing financial assistance to JSPL in making this expansion, but also providing them with green credibility in being at the forefront of the emerging carbon market.[14]

The CDM may even act as a disincentive for Southern governments considering climate-friendly legislation. Had it been mandatory for factories to capture and destroy HFC-23, they would not have qualified for CDM status, as the carbon funding would not have been considered to be "additional."

As far back as 1991, plans were drawn up for an EU-wide carbon tax, but the lack of political support and the vogue for all things market-related meant that they were pushed aside.[15] However, in February 2007 a study by economist Robert Shapiro, former undersecretary of commerce for economic affairs in the Clinton administration, stated that carbon taxes are "much less vulnerable to evasion and market manipulation" than cap-and-trade systems. Whereas carbon taxes provide "a more stable and transparent system for consumers and industry alike," cap-and-trade systems are "much more complex to administer" and "produce much greater volatility in energy and energy-related prices."[16]

Across the world, other economists and political scientists are coming to similar conclusions. The question remains how long so much energy and political will power will be channelled into a mechanism that does little more than bolster the profits and environmental credibility of the biggest polluters. Even if the global community will not have benefited from any serious net emissions reductions as a result of the EU-ETS, it will hopefully at least have learned a valuable lesson in how not to devise effective climate policy.

Community Interest

Nonetheless, there is gradually emerging a groundswell of opinion that the invisible hand of the market is not the most effective way of facing the climate challenge. The Durban Declaration of Climate Justice, signed by civil society organisations from all over the world, for example, asserts that making carbon a commodity represents a large-scale privatization of the Earth's carbon-cycling capacity, with the atmospheric pie having been carved-up and handed over to the biggest polluters.[17]

Effective action on climate change involves demanding, adopting and supporting policies that reduce emissions at source as opposed to offsetting or trading. Carbon trading isn't an effective response; emissions have to be reduced across the board without elaborate get-out clauses for the biggest polluters. There is an urgent need for stricter regulation, oversight, and penalties for polluters on community, local, national and international levels, as well as support for communities adversely impacted by climate change. But currently such policies are mostly invisible, as they contradict the sacred cows of economic growth and the free market. But unfortunately, when it comes to tackling climate change and maintaining an economic growth based on the ever-increasing extraction and consumption of fossil fuels, there is no win-win solution.

Market-based mechanisms such as carbon trading are an elaborate shell game of global creative accountancy. They distract people from the fact that no viable business-as-usual scenario exists. Climate policy needs to be made of sterner stuff.

Notes

1. Larry Summers, "The Memo," The Whirled Bank Group: A World Full of Poverty, May 11, 2007, http://www.whirledbank.org/ourwords/summers.html.
2. Nicholas Stern, "The Stern Review: A Summary," http://news.bbc.co.uk/1/hi/business/6098362.stm.
3. Tom Burke, "This Is Neither Scepticism Nor Science—Just Nonsense," The Guardian, October 23, 2004, http://education.guardian.co.uk/higher/research/story/0,,1334277,00.html
4. J. Kay, "Why the key to carbon trading is to keep it simple," Financial Times, May 9, 2006.
5. D. Gow, "Smoke alarm: EU shows carbon trading is not cutting emissions," The Guardian, April 3, 2007.
6. Kay, "Why the key to carbon trading."
7. The Durban Declaration on Carbon Trading, http://www.sinkswatch.org/pubs/Durban%20DeclarationSeptember%202006%20leaflet.pdf; "Hoodwinked in the Hothouse: The G8, Climate Change and Free-Market Environmentalism," http://www.carbontradewatch.org/pubs/hothousecolour.pdf.

8. K. Neuhoff, et al., "Comparison of National Allocation Plans for the Period 2008–2012," http://www.climate-strategies.org.
9. K. Neuhoff, et al., "Implications of Announced Phase 2 National Allocation Plans," *Climate Policy* 6 (2006): 416.
10. N. Wara, "Is the Global Carbon Market Working?" *Nature*, February 8, 2007, pp. 595–96.
11. Ibid.
12. S. Choudary, "SRF raises Rs 500 crore from carbon credits sale," *Economic Times of India*, April 9, 2007.
13. *Point Carbon*, "Indian chemical company books ₪87 million windfall from carbon trading," April 10, 2007.
14. Larry Lohmann, ed., *Carbon Trading? A Critical Conversation on Climate Change, Power and Privatisation* (The Dag Hammarskjold Foundation, 2006), Chapter 4, p. 259.
15. "Expert calls for EU-wide CO2 taxation," ENDS Europe DAILY 2290, March 26, 2007.
16. *Point Carbon*, "Study favours carbon tax over cap-and-trade," February 19, 2007.
17. The Durban Declaration on Carbon Trading.

The Corporate Climate Coup

by David Noble

Until quite recently most people were either unaware of or confused and relatively unconcerned about the climate change issue, despite a growing consensus among scientists and environmentalists about its possible dangers. Global warming activists, such as Al Gore, were quick to place the blame for that popular ignorance, confusion, and lack of concern on a well-financed corporate propaganda campaign by oil and gas companies and their front organizations, political cronies, advertising and public relations agencies and media minions, which lulled people into complacency by sowing doubt and skepticism about worrisome scientific claims. And, of course, they were right; there was such a corporate campaign, which has by now been amply documented.[1] What global warming activists conveniently failed to point out, however, is that their own, alarmist, message has been drummed into our minds by the very same means, albeit by different corporate hands. This campaign, which might well prove the far more significant, has heretofore received scant notice.

Over the last decade and a half we have been subjected to two competing corporate campaigns, echoing different time-honored corporate strategies and reflecting a split within elite circles. The issue of climate change has been framed from both sides of this elite divide, giving the appearance that there are only these two sides. The first campaign, which took shape in the late 1980s as part of the triumphalist "globalization" offensive, sought to confront speculation about climate change head-on by denying, doubting, deriding and dismissing distressing scientific claims which might put a damper on enthusiasm for expansive capitalist enterprise. It was modelled after and to some extent built upon the earlier campaign by the tobacco industry to sow skepticism about mounting evidence of the deleterious health-effects of smoking. In the wake of this "negative" propaganda effort, any and all critics of climate change and global warming have been immediately identified with this side of the debate.

The second positive campaign, which emerged a decade later, in the wake of Kyoto and at the height of the anti-globalization movement, sought to get out ahead of the environmental issue by affirming it—only to hijack it and turn it to corporate advantage. Modelled on a century of corporate liberal cooptation of popular reform movements and regulatory regimes, it aimed to appropriate the issue in order to moderate its political implications, thereby rendering it compatible with corporate economic, geopolitical, and ideological interests. The corporate climate campaign thus emphasized the primacy of "market-based" solutions. At the same time it hyped the global climate issue into an obsession, a totalistic preoccupation with which to divert attention from the radical challenges of the global justice movement.

The first campaign, dominant throughout the 1990s, suffered somewhat from exposure and became relatively moribund early in the Bush II era, albeit without losing influence within the White House (and the Prime Minister's Office). The second, having contributed to the diffusion of a radical movement, has succeeded in generating the current hysteria about global warming, now safely channeled into corporate-friendly agendas at the expense of any serious confrontations with corporate power. Its media success has aroused the electorate and compelled even die-hard deniers to disingenuously cultivate a greener image. Meanwhile, and most important, the two opposing campaigns have together effectively obliterated any space for rejecting them both.

The Climate Change Deniers

In the late 1980s the world's most powerful corporations launched their "globalization" revolution, incessantly invoking the inevitable beneficence of free trade and, in the process, relegating environmental issues to the margins and reducing the environmentalist movement to rearguard actions. Interest in climate change nevertheless continued to grow. In 1988, climate scientists and policymakers established the Intergovernmental Panel on Climate Change (IPPC) to keep abreast of the matter and issue periodic reports. At a meeting in Toronto three hundred scientists and policy-makers from forty-eight countries issued a call for action on the reduction of carbon dioxide emissions. The following year fifty oil, gas, coal and automobile and chemical manufacturing companies and their trade associations formed the Global Climate Coalition (GCC), with the help of public relations giant Burson-Marsteller. Its stated purpose was to sow doubt about scientific claims and forestall political efforts to reduce greenhouse-gas emissions. The GCC gave millions of dollars in political contributions and in support of a public relations campaign warning that misguided efforts to reduce greenhouse-gas emissions through restrictions on the burning of fossil fuels would undermine the promise of globalization and cause economic ruin. GCC efforts effectively put the climate change issue on hold.[2]

Meanwhile, following an indigenous uprising in Chiapas in January 1994 set for the first day of the implementation of the North American Free Trade Agreement the anti-globalization movement erupted in worldwide protest against market capitalism and corporate depredation, including the despoiling of the environment. Within five years the movement had grown in cohesion, numbers, momentum and militancy and coalesced in designated "global days of action" around the world. The campaign of direct actions at G8 summits and meetings of the World Bank, the International Monetary Fund and the World Trade Organization reached its peak in shutting down the WTO meetings in Seattle in November 1999. The movement, which consisted of a wide range of diverse grassroots organizations united in opposition to the global "corporate agenda," shook the elite globalization campaign to its roots. It was in this charged context that the signatories of the UN Framework Convention on Climate Change, which had been formulated by representatives from 155 nations at the Rio Earth Summit in 1992, met at the end of 1997 in Kyoto. There they established the so-called Kyoto Protocol to reduce greenhouse gas emissions through carbon targets and trading. The Kyoto treaty, belatedly ratified only in late 2004, was the sole international agreement on climate change and immediately became the bellwether of political debate about global warming.

Corporate opposition anticipated Kyoto. In the summer of 1997 the US Senate passed a unanimous resolution demanding that any such treaty must include the participation end compliance of developing countries, particularly emerging economic powers like China, India and Brazil, which were nevertheless excluded in the first round of the Kyoto Protocol. Corporate opponents of Kyoto in the GCC, with the swelling global justice movement as a backdrop, condemned the treaty as a "socialist" or "third-world" plot against the developed countries of the West.

The convergence of the global justice movement and Kyoto, however, prompted some of the elite to rethink and regroup, which created a split in corporate ranks regarding the issue of climate change. Defections from the GCC began in 1997 and within three years had come to include such major players as Dupont, BP, Shell, Ford, Daimler-Chrysler and Texaco. Among the last GCC hold-outs were Exxon, Mobil, Chevron and General Motors. (In 2000, the GCC finally went out of business but other like-minded corporate front organizations were created to carry on the "negative" campaign, which continues.)

The Climate Change Market-Makers

Those who split off from the GCC quickly coalesced in new organizations. Among the first of these was the Pew Center for Global Climate Change, funded by the philanthropic offering of the Sun Oil/Sunoco fortune. The board of the new center was chaired by Theodore Roosevelt IV, great grandson of the Progressive Era president (and conservation icon) and managing director of the Lehman Brothers investment banking firm. Joining him on the board were the managing director of the Castle-Harlan investment firm and the former CEO of Northeast Utilities, as well as veteran corporate lawyer Frank E. Loy, who had been the Clinton administration's chief negotiator on trade and climate change.

At its inception the Pew Center established the Business Environmental Leadership Council, chaired by Loy. Early council members included Sunoco, Dupont, Duke Energy, BP, Royal Dutch/Shell, Duke Energy, Ontario Power Generation, DTE (Detroit Edison) and Alcan. Marking their distance from the GCC, the council declared "we accept the views of most scientists that enough is known about the science and environmental impacts of climate change for us to take actions to address the consequences." The statement continued, "Businesses can and should take concrete steps now in the U.S. and abroad to assess opportunities for emission reductions ... and invest in new, more efficient products, practices, and technologies."[3]

The council emphasized that climate change should be dealt with through "market-based mechanisms" and by adopting "reasonable policies," and expressed the belief "that companies taking early action on climate strategies and policy will gain sustained competitive advantage over their peers."

Early in 2000, "world business leaders" convening at the World Economic Forum in Davos, Switzerland, declared that "climate change is the greatest threat facing the world." That fall, many of the same players, including Dupont, BP, Shell, Suncor, Alcan and Ontario Power Generation, as well as the French aluminum manufacturer Pechiney, joined forces with the US advocacy group Environmental Defense to form the Partnership for Climate Action. Like-minded Environmental Defense directors included the Pew Center's Frank Loy and principals from the Carlyle Group, Berkshire Partners and Morgan Stanley and the CEO of Carbon Investments.

Echoing the Pew Center mission, and barely a year after the "Battle of Seattle" had shut down the World Trade Organization, the new organization reaffirmed its belief in the beneficence of market capitalism. "The primary purpose of the Partnership is to champion market-based mechanisms as a means of achieving early and credible action on reducing greenhouse gas emissions that is efficient and cost-effective."[4]

Throughout its initial announcement this message was repeated like a mantra: "the benefits of market mechanisms," "market-oriented rules," "market-based programs can provide the means to simultaneously achieve both environmental protection and economic development goals," "the power of market mechanisms to contribute to climate change solutions." In the spring of 2002, the partnership's first report proudly stated that the companies of the PCA are in the vanguard of the new field of greenhouse-gas management. "The PCA is not only achieving real reductions in global warming emissions," the report noted, "but also providing a body of practical experience, demonstrating how to reduce pollution while continuing to profit."[5]

This potential for profit making from climate change gained the avid attention of investment bankers, some of whom were central participants in the PCA through their connections with the boards of the Pew Center and Environmental Defense. Goldman Sachs became the leader of the pack. With its ownership of power plants through Cogentrix and clients like BP and Shell, the Wall Street firm was most attuned to the opportunities. In 2004 the company began to explore the "market-making" possibilities and the following year established its Center for Environmental Markets. It announced that "Goldman Sachs will aggressively seek market-making and investment opportunities in environmental markets." The firm indicated that the center would engage in research to develop public policy options for establishing markets around climate change, including the design and promotion of regulatory solutions for reducing greenhouse gas emissions. The firm also indicated that Goldman Sachs would "take the lead in identifying investment opportunities in renewable energy." That year the investment banking firm acquired Horizon Wind Energy, invested in photovoltaics with Sun Edison, arranged financing for Northeast Biofuels and purchased a stake in Logen Corporation, which pioneered the conversion of straw, corn stalks and switch grass into ethanol. The company also dedicated itself "to act as a market maker in emissions trading" of carbon dioxide (and sulfur dioxide) as well as in such areas as "weather derivatives," "renewable energy credits" and other "climate-related commodities." "We believe," Goldman Sachs proclaimed, "that the management of risks and opportunities arising from climate change and its regulation will be particularly significant and will garner increasing attention from capital market participants."[6]

Enter Al Gore

Among those capital market participants was former US Vice-President Al Gore. Gore had a long-standing interest in environmental issues and had represented the US in Kyoto. He also had equally long-standing family ties with the energy industry through his father's friendship with Armand Hammer and his financial interest in Hammer's company Occidental Petroleum, which the son inherited. In 2004, as Goldman Sachs was gearing up its climate-change market-making initiatives in quest of green profits, Gore teamed up with Goldman Sachs executives David Blood, Peter Harris and Mark Ferguson to establish the London-based environment investment firm Generation Investment Management (GIM), with Gore and Blood at its helm.

In May 2005 Gore, representing GIM, addressed the Institutional Investor Summit on Climate Risk and emphasized the need for investors to think in the long term and to integrate environmental issues into their equity analyses. "I believe that integrating the issues relating to climate change into your analysis of what stocks are worth investing in, how much, and for how long, is simply good business," Gore explained to the assembled investors. Applauding a decision to move in this direction announced the day before by General Electric's CEO Jeff Immelt, Gore declared that "We are here at an extraordinarily hopeful moment... when the leaders in the business sector begin to make their moves." By that time Gore was already at work on his book about global warming, *An Inconvenient Truth*, and that same spring he began preparations to make a film about it.[7]

The book and the film of the same name both appeared in 2006, with enormous promotion and immediate success in the corporate entertainment industry (the film eventually garnering an Academy Award). Both vehicles vastly extended the reach of the climate-change market-makers, whose efforts they explicitly extolled. "More and more U.S. business executives are beginning to lead us in the right direction," Gore exulted, adding "there is also a big change underway in the investment community."

The book and film faithfully reflected and magnified the central messages of the corporate campaign. Like his colleagues at the Pew Center and the Partnership for Climate Action, Gore stressed the importance of using market mechanisms to meet the challenge of global warming. "One of the keys to solving the climate crisis," he wrote, "involves finding ways to use the powerful force of market capitalism as an ally." Gore repeated his admonition to investors about the need for long-term investment strategies and for integrating environmental factors into business calculations, proudly pointing out how business leaders had begun "taking a broader view of how business can sustain their profitability over time." The one corporate executive actually quoted in the book, in a two-page spread, was General Electric's CEO Jeffrey Immelt, who succinctly explained the timing and overriding purpose of the exercise: "This is a time period where environmental improvement is going to lead to profitability."[8]

A Corporate Call for Action

By the beginning of 2007 the corporate campaign had significantly scaled up its activity, with the creation of several new organizations. The Pew Center and Partnership for Climate Action now created a political lobbying entity, the US Climate Action Partnership (USCAP). USCAP membership included the key players in the initial effort, such as BP, Dupont, the Pew Center and Environmental Defense, and added others, including GE, Alcoa, Caterpillar, Duke Energy, Pacific Gas and Electric, Florida Power and Light and PNM, the New Mexico and Texas utilities holding company. PNM had recently joined with Microsoft's BUI Gates' Cascade Investments to form a new unregulated energy company focussed on growth opportunities in Texas and the western US. PNM's CEO Jeff Sterba also chaired the Climate Change Task Force of the Edison Electric Institute. Also joining USCAP was the Natural Resources Defense Council, the World Resources Institute and the investment banking firm Lehman Brothers whose managing director Theodore Roosevelt IV chaired the board of the Pew Center and was soon also to chair Lehman's new Global Center on Climate Change. As *Newsweek* noted, "Wall Street is experiencing a climate change," with the recognition that "the way to get the green is to go green."[9]

In January 2007, USCAP issued "A Call for Action," a "non-partisan effort driven by the top executives from member organizations." The call declared the "urgent need for a policy framework on climate change," stressing that "a mandatory system is needed that sets clear, predictable, market-based requirements to reduce greenhouse gas emissions."[10]

USCAP carved out a "blueprint for a mandatory economy-wide market-driven approach to climate protection," which recommended a "cap and trade" program as its cornerstone, combining the setting of targets with a global carbon market for trading emission allowances and credits.[11] Long condemned by developing countries as "carbon colonialism," carbon trading had become the new orthodoxy. The blueprint also called for a "national program to accelerate technology, research, development, and deployment and measures to encourage the participation of developing countries like China, India, and Brazil, insisting that "ultimately the solution must be global." According to USCAP spokesperson General Electric's CEO Jeff Immelt, "these recommendations should catalyze legislative action that encourages innovation and fosters economic growth while enhancing energy security and balance of trade."[12]

The following month yet another corporate climate organization made its appearance, this one specifically dedicated to spreading the new global warming gospel. Chaired by Al Gore of Generation Investment Management, the Alliance for Climate Protection included among its members the now familiar Theodore Roosevelt IV from Lehman Brothers and the Pew Center, former national security advisor Brent Scowcroft, Owen Kramer from Boston Provident, representatives from Environmental Defense, the Natural Resources Defense Council and the National Wildlife Federation, and three former Environmental Protection Agency Administrators. Using "innovative and far-reaching communication techniques," Gore explained, "the Alliance for Climate Protection is undertaking an unprecedented mass persuasion exercise"—the multimedia campaign against global warming now saturating our senses.

A Strategy to Distract

If the corporate climate-change campaign has fuelled a fevered popular preoccupation with global warming, it has also accomplished much more. Having arisen in the midst of the worldwide global justice movement, it has restored confidence in those very faiths and forces that that movement had worked so hard to expose and challenge. These include globe-straddling profit-maximizing corporations and their myriad agencies and agendas; the unquestioned authority of science and the corollary belief in deliverance through technology; and the beneficence of the self-regulating market with its panacea of prosperity through free trade, and its magical powers which transform into commodities all that it touches, even life. All the glaring truths revealed by that movement about the injustices, injuries and inequalities sowed and sustained by these powers and beliefs have now been buried, brushed aside in the apocalyptic rush to fight global warming.

Explicitly likened to a war, this epic challenge requires single-minded attention and total commitment, without any such distractions. Now is not the time, nor is there any need, to question a deformed society or re-examine its underlying myths. The blame and the burden has been shifted back again to the individual, awash in primordial guilt, the familiar sinner facing punishment for his sins, his excesses, predisposed by his pious culture and primed now for discipline and sacrifice.

On opening day of the 2007 baseball season, the owner of the Toronto Blue Jays stood in front of the giant jumbotron, an electronic extravaganza, encircled by a ring of dancing corporate logos and advertising, and exhorted every person in the crowd, preposterously, to go out and buy an energy-efficient light bulb. They applauded.

Notes

1. See for example George Monbiot, *Heat, How to Stop the Planet From Burning* (Doubleday, 2006), Chapter 2.
2. *Los Angeles Times*, December 7, 1997. See also the entries for "Global Climate Change" at http://www.sourcewatch.com and http://www.commoncause.com.
3. Vicki Arroyo Cochran, Director of Policy Analysis, Pew Center on Global Climate Change, "Legislation Needed—Before It's Too Late," *Washington Post*, Special Section on Climate Change, Opinion Editorial, October 25, 1999, http://www.pewclimate.org/press_room/opinion_editorials/oped_earlyaction.cfm.
4. Environmental Defense, "Global Corporations and Environmental Defense Partner to Reduce Greenhouse Gas Emissions Partnership Pioneers Real-World Solutions," October 17, 2000, http://www.environmentaldefense.org/article.cfm?ContentID=503.
5. Environmental Defense, "Partnership For Climate Action Releases Report On Corporate Greenhouse Gas Management Programs," April 2, 2002, http://www.environmentaldefense.org/pressrelease.cfm?ContentID=1886.
6. Goldman Sachs, Goldman Sachs Environmental Policy Framework, http://www2.gold-

mansachs.com/our_firm/our_culture/corporate_citizenship/environmental_policy_frame-work/docs/EnvironmentalPolicyFramework.pdf.

7. Al Gore, *An Inconvenient Truth* (Rodale Books, 2006).
8. Johnnie L. Roberts, "The Street Turns Green," *Newsweek*, March 12, 2007.
9. "A Coalition for Firm Limits on Emissions," *New York Times*, January 19, 2007
10. "Big Business Pushes Bush on Carbon Caps," Columbia Broadcasting System Interactive Inc., January 23, 2007.
11. "Major Business and Environment Leaders Unite to Call for Swift Action on Global Climate Change," National Resource Defence Council press release, January 22, 2007.
12. "G.E. Chief Urges U.S. to Adopt Clear Energy Policy," *New York Times*, May 10, 2005.

Turning on Canada's Tap?

by Tony Clarke

When Prime Minister Stephen Harper sat down with President George
W. Bush in their first White House meeting July 6, 2006, one of the
"unmentionable" items on their agenda may well have been the question
of bulk water exports from Canada. After all, Bush himself raised the issue
back in July 2001 when he talked off the cuff to reporters about growing
water shortages in his home state of Texas and elsewhere in the country,
saying he would like to begin negotiations with Ottawa on water exports
from Canada. In Texas, he said, "water is more valuable than oil." "A lot of
people don't need it, but when you head south and west, we need it," Bush
declared, adding that he "looked forward" to discussing the matter with then
Prime Minister Jean Chretien.

At the time, the reaction from Canadian officials was swift and blunt.
"We're absolutely not going to export water, period," proclaimed David
Anderson, then Canada's environment minister. Anderson's comment reflected what seems to be a general public consensus that water should not
be treated like other natural resources (e.g., oil, gas, minerals, timber etc.) as
a commodity to be bought and sold on the open market to US customers.
After Anderson's reaction, the issue seemed to fade from the news headlines
until former US ambassador Paul Celucci revived the issue in the early stages
of the 2005–6 federal election.

The question now is whether Stephen Harper is willing to put water on
the table in negotiating a new relationship with the United States. Although
Harper's specific views on water exports are not known, he has called for more
"economic and security integration" with the US, highlighting the need for
a continental energy strategy that would include "a range of other natural
resources." As a new era of Canada-US relations opens up, Canadians would
do well to take a closer look at the forces moving behind the scenes to turn
on the taps for massive water exports to the United States.

Growing American Thirst

Today, the largest world's largest economic and military superpower is facing
the problem of acute water shortages within its own borders. Twenty-one
percent of farmland irrigation in the US comes from the pumping of groundwater at rates that exceed the water's ability to recharge. In effect, this means
those aquifers that are the country's source of freshwater are rapidly being

depleted and are dying up.[1] The lethal combination of severe droughts and dried up wells has become the plague of many US farmers. Every year now, it is estimated that more than US$400 billion is lost in America's farmlands because of the depletion of aquifers. A prime example is the Ogallala aquifer, one of the world's most famous underground bodies of water which is being depleted at a rate of fourteen times faster than nature can restore it. It is now estimated that more than half of the water is already gone.[2]

In California, the major aquifers are also drying up. With the Colorado River strained to the limit, the water table under California's San Joaquin Valley has dropped nearly ten meters in some areas during the past fifty years. In the state's Central Valley, overuse of underground water supplies has resulted in a loss of over 40 percent of the combined storage capacity of all the human made reservoirs in California.[3] The desert regions of the American southwest—Arizona, Nevada and New Mexico—largely barren of water, continue to experience population growth. In Bush's home state of Texas, water scarcity is also approaching a critical stage where cities like El Paso are expected to be dried up by 2030. Moving further into the American Midwest, Chicago and Milwaukee could be facing severe water shortages. The huge sandstone aquifer underlying the Illinois-Wisconsin border, which supplies these two major cities with their water supplies, is currently overtaxed and may well be depleted in the near future, say scientists, unless there are significant reductions in groundwater withdrawals.

In short, the US is becoming more and more thirsty as it reaches the danger point of running out of its own freshwater sources. When the US government surveyed the fifty states of the union in 2003, it found that more than two-thirds predicted they would face water shortages in one form or another over the next ten years.[4] Nor does the US government seem prepared it confront this impending water crisis. In June 2004, the National Academies of Science (NAS) and the US Geological Survey reported that Washington is ill-prepared to deal with water shortages emerging across the country. NAS reports have also shown that overall federal funding for water research in the US has remained stagnant for the past thirty years.[5]

Water-Rich Canada?

As the US water crisis intensifies, one quick fix solution is to tap into what is perceived to be Canada's considerable water wealth. According to this scenario, Canada is a giant green sponge full of freshwater lakes and rivers—a massive reservoir of water that can be tapped to serve the insatiable thirst of people and industries in urban America. Globally speaking, Canada is ranked fourth in the world in terms of surface sources of freshwater in the form of lakes, rivers and glaciers.[6] All it would take is the construction of a network of new dams, reservoirs, canals, tunnels, pipelines and supertankers to transport water in bulk form from Canada to the United States.

The largest freshwater system on the planet is, of course, the Great Lakes lying between Canada and the United States, which contain no less than 20 percent of the world's freshwater. But the Great Lakes have also been a dumping ground for industrial wastes, contaminating much of the lake water and ground water in the region. Indeed, the International Joint Commission declared in its 2000 "Final Report on the Protection of Waters in the Great lakes" that there is no surplus water in the Great Lakes and emphatically warned against any new diversions. Moreover, scientists are now warning that drought patterns are returning to the Prairies as river systems like the South Saskatchewan, Old Man, Peace and Athabasca show signs of drying up. And these re-emerging drought patterns are bound to intensify with global warming.[7] Already, the glacier that feeds the Bow River in Alberta is melting so quickly that there may be no water left in it fifty years from now.

Although there is no doubt that Canada is blessed by nature's endowment with numerous freshwater lakes and rivers, it should also be noted that 60 percent of our rivers flow north into the Hudson Bay and the Arctic regions. As a result, 60 percent of Canada's freshwater flows in the opposite direction of the US and is largely inaccessible. Even so, say politicians, engineers and economists on both sides of the border, there are ways of overcoming these obstacles through new technology and investment.[8]

Mega Export Schemes

Over a past four decades, a series of mega diversion schemes have been planned for massive bulk water transfers from Canada to the US. In retrospect, these mega projects can be categorized in terms of three major water corridors.[9]

Western Corridor. The centerpiece of the western water corridor flowing from Canada to the US is the North American Water and Power Alliance. NAWAPA was originally designed to bring bulk water from Alaska and northern British Columbia for delivery to thirty-five US States. By building a series of large dams, the northward flow of the Yukon, Peace, Liard and a host of other rivers (Tanana, Copper, Skeena, Bella Coola, Dean, Chilcotin and Fraser) would be reversed to move southward and pumped into the Rocky Mountain Trench where the water would be trapped in a giant reservoir approximately 800 kilometers long. A canal would then be built to take the water southward into Washington state where it would be channeled through existing canals and pipelines to supply freshwater for customers in thirty-five states. The annual volume of water to be diverted through the NAWAPA project is estimated to be roughly equivalent to the average total yearly discharge of the entire St. Lawrence River system in Canada.

Central Corridor: Another water corridor consists of a series water diversions schemes proposed from the Northwest Territories through the Prairies into the US. In 1968, the Washington State Resource Center developed plans for the Central North American Water Project (CeNAWAP). The plan calls for a series of canals and pumping stations linking Great Bear Lake and Great Slave Lake in the Northwest Territories to Lake Athabaska and Lake Winnipeg and then the Great Lakes for bulk water exports to the US. A variation on the CeNAWAP is the Kuiper Diversion Scheme, which proposes to link the major western rivers into a mega water diversion scheme involving the Mackenzie, the Peace, the Athabasca, North Saskatchewan, Nelson and Churchill river systems. Both the Kuiper and the CeNAWAP are designed to transport water for export to the US.

Eastern Corridor: The principal eastern water corridor is known as the GRAND Canal—the Great Recycling and Northern Development Canal. As originally conceived, the GRAND Canal plans call for the damming and rerouting of northern river systems in Quebec in order to bring freshwater through canals down into the Great Lakes where it would be flushed into the American Midwest. A dike would be built across James Bay at the mouth of Hudson Bay (whose natural flow is northward), thereby turning the bay into a giant fresh water reservoir of about thirty thousand square miles supplied by twenty rivers. Through a system of dikes, canals, dams, power plants and locks, the water would then be diverted from the reservoir and rerouted southward down a 167 mile canal at a rate of about 282,000 liters a second into two of the Great Lakes—Lake Superior and Lake Huron. From there, the water would be flushed through canals into markets in both the American Midwest and US Sun Belt.

There are, of course, multiple reasons why none of these massive water corridors have been built in the intervening decades since they were first proposed.[10] One reason is that America's thirst has been temporarily quenched by internal bulk water transfers within the US. A second reason is the problem of securing sufficient capital to pay for highly expensive bulk water export schemes like those described above and whether this should be private or public investment. A third reason may also have been the need for new or improved engineering technologies required for some of the more geographically challenging projects.

Yet, underlying all these reasons is the question of political will. According to opinion polls, most Canadians remain skeptical about selling our water to the US. In a 2002 survey conducted by the Centre for Research and Information on Canada, 69 percent were opposed to bulk water exports. Three years before this poll was taken, the House of Commons actually passed a motion (introduced by the New Democrats) calling on the federal government to ban the export of water. In response, the Liberal government of the day refused to issue a federal ban on water exports contending it would contravene Canada's obligations under NAFTA, and, instead, worked with the provinces to develop a Canada-wide water accord aimed at discouraging bulk-water exports. Yet, precisely because of NAFTA, which prohibits countries from putting a ban or quota on the exports of their natural resources, the water accord is largely ineffective.

Which brings us back to Stephen Harper. Will he and his government be the ones who finally muster the political will to give Washington the green light and permit bulk-water takings? Is this the price that Harper is prepared to pay, on behalf of Canadians, to seal a new grand bargain with the US just as Brian Mulroney did when he gave away Canada's energy resources in the eleventh hour of the free trade negotiations?

Notes

1. See analysis of US water shortages in Maude Barlow and Tony Clarke, *Blue Gold: The Battle Against the Corporate Theft of the World's Water* (Toronto: McClelland and Stewart, 2002), Chapter 1.
2. Ibid.
3. Ibid.
4. Cited in Maude Barlow, *Too Close For Comfort: Canada's Future Within Fortress North America* (Toronto: McClelland and Stewart, 2005), pp. 215ff.
5. Ibid.
6. Barlow and Clarke, *Blue Gold*, p. 151. The reference here is to surface water studies by Russian hydrologist Igor Shiklomanov. Other studies rank Canada third to sixth in terms of renewable water supplies. See John B. Sprague, "Great White North, Canada's myth of Water Abundance," in Karen Baker, ed., *Eau Canada: The Future of Canada's Water* (Vancouver: UBC Press, 2007), pp. 23–36.
7. See the work of renowned Canadian water scientist David Schindler from the University of Alberta. For example, David Schindler, "The Cumulative Effects of Climate Warming and Other Human Stresses on Canadian freshwaters in the new Millennium," *Canadian Journal of Fisheries and Aquatic Sciences* 58: 18–29.
8. For some historical background, see Richard Bocking, *Canada's Water: For Sale?* (Toronto: James Lewis and Samuel, 1972).
9. Although the naming and framing of these three water export corridors is my undertaking, the research is outlined in Frederic Lasserre, "Les Projets de Transferts Massifs d'Eau en Amerique du Nord," *Vertigo*, hors-serie no. 1.
10. See also Frederic Lasserre, "Drawers of Water: Water Diversions in Canada and Beyond," in Baker, ed., *Eau Canada*, pp. 143–62.

Fudging the Numbers

Stephen Harper's Response to Climate Change

by Dale Marshall

Over the course of 2006, there was a significant surge in attention and concern being paid to climate change in Canada and abroad. The reason for this is debatable—more compelling science, more media attention or penetration of the issue into popular culture.

Whatever the reason, this was a global phenomenon. The epicenter was arguably the United States. Former US vice-president Al Gore's documentary movie *An Inconvenient Truth* gained so much attention throughout the year that it barely benefited from increased profile when the film won an Oscar the next year. Hurricane Katrina was also a landmark event, both due to its devastation and because, for the first time, it provoked a debate in the media and the general public about the links between extreme weather and climate change.

But governments also helped to raise the profile of climate change. The UK, as host of the G8 in 2005, made it one of two major issues to be considered at the meeting in Gleneagles. The UK also commissioned a comprehensive report on the economics of climate change released by Sir Nicholas Stern in October 2006. It was impossible to dismiss Stern as alarmist when the former World Bank's chief economist concluded that the costs of inaction on climate change were five to twenty times greater than the cost of tackling it head on.[1] In Canada, all three opposition parties questioned the government repeatedly on its climate-change plans in the House of Commons. Canada's environment commissioner evaluated Canada's record on climate change (largely a Liberal one) as poor and called for a "massive scale-up of effort" to address the issue.[2]

The scientific evidence of human-caused climate change, though hard to ignore previously, nevertheless developed a newfound persuasiveness. So that by the first half of 2007, when the Intergovernmental Panel on Climate Change (IPCC) released its *Fourth Assessment Report* calling human interference with the climate system "unequivocal," each of the three-staged reports drew more attention than the previous.[3]

None of the science or even the economic analysis was terribly new. The scientific community had been publishing dire warnings about the potential impacts of climate change for almost two decades.[4] The cost of addressing climate change had been investigated in numerous studies before Stern and found to be modest and manageable.[5]

What was new and likely an important determining factor in rising public concern was the media's treatment of climate change. Whether it was the front page of newspapers, the lead story on evening newscasts or the subject of talk-radio debate, the media was paying attention like it never had before. Climate change and global warming also penetrated popular culture, the subject or theme of numerous movies (not just Mr. Gore's), high circulation magazines, and even ad campaigns.

Canada's new government, however, has not reflected shifting public opinion with adequate climate-change policies. The government has, in fact, profoundly misread the extent to which concern about climate change *has* penetrated the Canadian consciousness, and the scale-up of effort requested by Canada's environment commissioner has not been seen. Ironically, the considerable gap between public concern and political action likely helped to vault the issue to the top of the agenda during the Conservatives' first year in power. By the end of the year, with polls showing that the environment, specifically climate change, was the top concern of Canadians, Prime Minister Harper's approach shifted significantly.[6]

Year 1: Climate Change Concern Rises

The UN climate-change conference that took place in Montreal in December 2005 was a clear success. One hundred and sixty countries—the parties to the Kyoto Protocol—came together and agreed to negotiate a second phase of Kyoto that would immediately follow the first, in 2013. Unfortunately UN momentum took a sharp downturn the next year. Canada had assumed the presidency of the UN negotiations for one year, starting in Montreal. The role of president is to facilitate negotiations not only at the formal meetings but also through inter-sessional negotiations and informal bilateral meetings. It is through close attention to these types of negotiations that momentum is maintained and agreement could be reached on a path forward.

These negotiations are critical. To avoid dangerous climate change, the global community would have to negotiate and implement many more agreements that are on the same scale as Kyoto. In the second phase, developed countries would have to take on deeper emission reductions while larger developing countries would have to begin accepting commitments as well.

In January 2006 after the federal election, the Canadian presidency shifted to the new minister of environment in the Conservative government, Rona Ambrose. However for most of the year the president's chair sat empty and little direction was offered by the new president to move international negotiations forward.

Canadian embarrassment began in Bonn in May 2006. Every year, the German city plays host to an important mid-year negotiating and discussion session on the Kyoto Protocol. Environment Minister Rona Ambrose only showed up long enough to open the two-week session and then jetted back to Canada the next day. Worse, leaked instructions from the Harper administration to their negotiators—senior bureaucrats from Environment, Foreign Affairs and Natural Resources Canada—told them to block any progress in the talks. The instructions stated that "Canada will not support [an] agreement ... that commits developed countries to more stringent targets in the future."[7] They also urged a delay to negotiations by suggesting a two-year assessment period for the international community.

Minister Ambrose ignored other opportunities for climate-change diplomatic interventions and negotiations. She failed to show up to the Gleneagles Dialogue on Climate Change, an initiative that originated at the G8 summit in the summer and continued in Monterrey, Mexico, in the fall. A UN-organized ministerial climate-change meeting in Kapama Lodge, South Africa also failed to obtain her participation. International meetings intended to prepare for the year-end UN conference, this time in Nairobi, were held without the Canadian president.

Finally, when the annual UN negotiations opened in November in Nairobi, the minister appeared by videoconference to open the session. Her first responsibility was to transfer the presidency role from Canada to the host Kenyans.

The next week, Minister Ambrose formally addressed the international delegates and took the opportunity to blame the previous government for Canada's poor record living up to international obligations on climate change. Bringing up domestic politics is considered a *faux pas* in international diplomatic circles, and negotiators in the audience were described as surprised, disappointed and even embarrassed.[8]

Canada's performance as chair of and participant in the climate-change discussions over the course of 2006 drew considerable criticism from the international community. Often in the diplomatic world of international negotiations, criticism comes in soft rebukes veiled in code and wrapped in flowery niceties, but this was not always the case for barbs directed at Canada. In response to Minister Ambrose's saying Canada would not meet its Kyoto targets, the German environment minister said that Canada was not showing the same leadership it showed in the past, and that Canada could still meet its Kyoto targets.[9]

By the end of the year, the criticism became even sharper. The French environment minister said he was "extremely disappointed" that the Canadian government was abandoning Kyoto: "It's a shock for us, it's a shock for all those who signed on to Kyoto, but it's mostly a shock for Canadians today, who seem to be leaning more towards Kyoto than anything else."[10] The EU environment commissioner said that Canada not meeting its Kyoto commitments "would be dealing a damaging blow to international law and to its credibility in future negotiations."[11] A Tuvalu ambassador and vice-chair of the forty-three-member Alliance of Small Island States, said, "Canada criticizes other countries about their human rights policies or about the death penalty while they are playing with the lives of island people and the Inuit."[12]

The new federal government's domestic approach to climate change was just as dismissive. Within one month of their election win, climate-change programs were cut, first at Natural Resources Canada then at Environment Canada.

By the time the federal budget was released the trend was clear and, as expected, $3 billion in climate change programming was cut. The $1 billion Climate Fund, expected to deliver as much as 50 to 100Mt of greenhouse gas (GHG) emission reductions, was transferred to a tax-credit program for transit users and, according to Environment Canada's own numbers, would reduce emissions by less than 1Mt.[13] The budget speech delivered by the minister of finance in Parliament referenced a $2 billion climate-change plan, down from the previous government's $10 billion commitment, but even that allocation was not in the actual budget.[14]

With climate programs largely cut and the previous climate-change plan abandoned, what would the new government's approach to tackling climate change be? Throughout the spring, summer, and fall Prime Minister Harper and Environment Minister Ambrose kept telling Canadians to be patient, that a climate-change plan was forthcoming. In October, Mr. Harper flew himself, three cabinet ministers and their entourage to Vancouver to give small hints, but no details, on their plan. A week later, the government finally tabled the *Clean Air Act*.[15] The only substantive measure on climate change was a 2050 target for Canada's greenhouse gas emissions. Commentators pointed out that Stephen Harper and his environment minister would be octogenarians before they could be held to account.

Within hours, it was clear that the *Clean Air Act* would not be passed as drafted. The three opposition parties immediately panned it, saying they would vote against it. Canadians would be forgiven for thinking that the government initiative was dead and that they would feel pressure to come back with something stronger.

But NDP leader Jack Layton relieved some of that pressure by recommending that the Act be sent to a special all-party committee for consideration. Because this would happen before its second reading in the House, the Act could be heavily amended. The Prime Minister quickly agreed. Environmental groups were worried about this deal because the issue could disappear in a House of Commons committee, often where public scrutiny goes to die.

Five months later, the heavily amended bill re-emerged as a serious plan to tackle greenhouse-gas emissions. The new bill, now called the *Clean Air and Climate Change Act*, had Kyoto-level targets for industry up to 2012, much stronger GHG emission reduction targets for 2020 and 2050 and good air quality provisions.[16] It was not perfect—industry could pay into an investment fund instead of reducing emissions, at the cost of $20 to 30 per tonne of carbon dioxide emitted. But many of the recommendations made to the committee by Canadian environmental groups were incorporated and so the amended Act was considered a vast improvement in terms of addressing climate change.

Prime Minister Harper's government did not support the changes. After the amendments were finalized, Brian Jean, a Conservative MP from Fort MacMurray, Alberta, walked away from a cake-cutting ceremony, saying, "As a personal position, I refuse to have cake at funerals."[17] Environment Minister John Baird, in media interviews, said he was "not happy," saying that the committee had "weakened" the original Act.[18] The government house leader, Peter Van Loan, confirmed a month later that the government had "no intention" of passing the heavily amended Act.[19]

Tackling Climate Change

Tackling climate change is not a simple matter. It requires many different types of policies that address all the major emitting sectors and includes a program to engage the general public. But it is reasonably easy to figure out if the overall plan is working; one simply has to look at one number—total greenhouse gas emissions—and determine if it is going up or down.

The science is clear that to avoid the planet warming by an average two degrees Celsius, a level widely understood to lead to "dangerous" climate changes, humans would have to cut emissions at least in half by mid-century. Canada is a rich developed country whose emissions remain one of the highest in the world on a per capita basis.[20] Canada must therefore reduce its emissions by 80 percent below 1990 levels by 2050 if it is to do its fair share.[21] So total GHG emissions need to start going down soon and continue to drop until we reach deep reductions. Globally, if the number that represents GHG emissions either stabilizes or, worse even, continues to climb, the world will not avoid dangerous climate-change impacts. Instead, the world will warm by more than of two degrees Celsius, a large proportion (possibly a majority) of the world's species will be at risk of extinction, significant portions of the Greenland and West Antarctic ice caps will melt, contributing to devastating sea-level rise, and these impacts and others will transform many millions of the world's citizens into climate refugees.[22]

So how does a country make its number—total GHG emissions—go down? First, there needs to be a systematic and comprehensive plan to address emissions from all sources: industry (the electricity, petroleum and manufacturing sector), transportation (by land, sea and air for both personal and goods movement), all buildings, landfills, deforestation and agricultural soils. It takes a combination of:

- Regulations, such as mandatory and continuous improvements in the efficiency of vehicles, buildings and appliances and anti-sprawl legislation;
- Financial penalties for polluting activities through a carbon tax or a cap-and-trade system for large industry;
- Funding programs for public transit and other sustainable options that are best delivered by government;
- Incentives for retrofitting buildings and other activities that are next to impossible to deliver otherwise; and
- Public engagement to create a collective vision for the solutions that are necessary.

Then, of course, the programs that make up the plan need to be put into place and adjusted as lessons are learned and improvements identified. In the end, it is easy to evaluate overall progress, by taking a look at that total emissions number.

You can fudge the accountability of this climate math by playing games with the numbers. One way to do this is by changing the baseline year. For fifteen years, the world has used 1990 as the base year from which percentage improvements need to be made. Canada's Kyoto target is 6 percent below 1990 emission levels for the period between 2008 and 2012. If emissions go up, and they have since 1990 in Canada, then changing the base year from say 1990 to 2006 makes the same percentage decrease in emissions easier, because emissions increases between the two dates are essentially ignored.

Another way to fudge the math is to use intensity-based GHG emission targets. The intensity of GHG emissions is the emissions based on the level of economic activity—a ratio of emissions per GDP or per barrel of oil for example. Setting intensity targets means only this ratio goes down, while overall GHG emissions would continue to rise. So, for example, a 17 percent improvement in GHG emissions intensity for the U.S. is actually a 12 percent *increase* in GHG emissions. A 50 percent decrease in emissions intensity in Alberta translates into a 33 percent *increase* in emissions.[23] Since the atmosphere does not respond to intensity, but rather to actual emissions, an intensity approach fails to address the problem.

The current federal government has served up both kinds of fudge. It has always insisted that, much like President Bush and Premier Klein, an intensity-based approach would work, and used the fuzzy math in both the *Clean Air Act* and industrial regulations. In the *Clean Air Act*, the government changed the base year from 1990 to 2003, making the 2050 target seem more impressive. In 2007, they changed the baseline again, to 2006.

The Liberal Record on Climate Change

The other main tactic that the federal Conservative government has used to deflect criticism on climate change is to point the finger at the previous Liberal government. "Thirteen years of Liberal inaction" must appear hundreds of times in House of Commons transcripts.

The Conservatives have a point. GHG emissions rose steadily during four Liberal governments and are now close to 30 percent above 1990 levels.[24] The Liberals, especially in their later years, did spend a considerable amount of money on climate change, but they relied too heavily on voluntary programs and incentives, and many programs that were developed were later seen to be ineffective. They eschewed almost entirely any regulations or placing a financial penalty on GHG emissions.

Stéphane Dion's record as environment minister is more mixed. Minister Dion released a climate-change plan in April 2005 whose biggest flaw was a very weak target for the Canadian industrial sector.[25] Though responsible for close to half of Canada's GHG emissions, the industrial target was only 13 percent of the overall. There was also a loophole introduced in the form of a technology fund that would further weaken GHG reductions. Industrial targets were intensity-based, and government documents showed that emissions from Canadian industry would continue to rise.[26]

The Liberal plan was a notable setback for climate protection. This was the first use of intensity-based targets at the federal level in Canada and helped to legitimize them. Other weaknesses, like the technology fund, also persist in Canadian policy.

The David Suzuki Foundation and ten other environmental groups called the plan "inadequate" to reach our Kyoto targets.[27] They recommended seven significant things the government could do to improve it, including strengthening the target for Canadian industry, establishing targets and timelines for each of the programs, and setting up a system of accountability and transparency. By the end of the year, most recommendations had gone unheeded. The Liberal lost the election the next January, though climate change and the environment were not a significant election issue.

On the positive side was Minister Dion's performance on the international scene. Just before the Liberals lost the election, at the UN Montreal climate-change conference, all countries that were part of the Kyoto Protocol were required to start negotiating the next phase of the Protocol, starting after 2012. There were significant splits between developed and developing countries and even within those camps on whether there should be another negotiated deal and, if so, who would take on what commitments. Mr. Dion, as chair of the conference, got all countries to agree to negotiate a treaty that would constitute the next phase of Kyoto. It was the culmination of a year of diligent work by Minister Dion, preparing the groundwork for the conference by attending and engaging in numerous multilateral and bilateral meetings on climate throughout the year. Though the details are still being negotiated and the discussions could still be derailed, the Montreal conference was a clear victory for those wanting global action on climate change.

Year 2: Playing with Numbers

The year 2007 brought a significant shift in Prime Minister Harper's approach to climate change, kicked off by a cabinet shuffle that placed John Baird in the environment portfolio.

For the next few months, the Conservative strategy had two parts: continue to accuse the former Liberal government of failing on the environment and adopting many preexisting Liberal programs. The names of the programs were rebranded as initiatives of the new federal government: ecoEnergy, ecoTransport, ecoTrust, etc.

The good news was that the government was at least moving in the right direction, reinstating climate-change programs and funding rather than cutting them as they had throughout their first year in office. The 2007 budget had $3 billion in "new" climate-change funding, though this was slightly less than what had been cut the year before.[28] Some good programs were put back into place, most notably a subsidy program for wind and other renewable energy sources, a grant program to entice people to retrofit their homes and money for provincial climate-change plans and programs.

The bad news was that the programs were reinstated in piecemeal fashion and with less money. The government also started funding headline grabbing but misguided projects such as a hydrogen highway in British Columbia and a carbon dioxide pipeline in Alberta. How much would emissions go down from these multi-million-dollar public investments? The government did not say and no analysis of the GHG benefits was released publicly.

It was photo-op politics, a communications strategy that involved numerous announcements but no comprehensive plan or a systematic approach to addressing climate change. The Conservatives, both in opposition and in government, accused the Liberals of spending lots of money on climate change without any accountability and little result. Now they were adopting the strategy wholesale.

With the completely amended *Clean Air Act* in a form they no longer supported, the government decided to move in a different direction. As environmental groups and the opposition had pointed out, they could simply add regulations to the Canadian *Environmental Protection Act* and, in April 2007, that is what they did.

The week before the big announcement, Environment Minister John Baird tried to temper expectations by releasing a study showing that Canada could only meet its Kyoto targets by provoking a recession, using faulty assumptions that artificially inflated Kyoto's economic costs.[29] Finally, when the Minister did release his plan, the GHG targets were a disappointment. Canada's emissions were expected to rise for three to five years and would still be 11 percent above Kyoto Protocol targets in 2020, eight years after the initial commitment period for the Protocol ends. With the science showing that Canada needed to be 25 percent below 1990 levels by 2020 and world leaders pledging to be 20 to 40 percent below, Canada's emissions would still be above 1990 levels in 2020.

Of course, that's not what the plan said. Minister Baird used fuzzy math to show emissions would improve by 20 percent or 38 percent, depending on whether he was using the changed baseline target or the intensity-based one.[30]

The details of the plan were even worse. The intensity-based target was just the first of the loopholes that make it unlikely that even the weak absolute targets would be reached. The regulations will also allow companies to reach up to 70 percent of their target—gradually decreasing over time—by simply paying fifteen dollars per tonne of carbon dioxide into a technology fund, an amount too low for most companies to contemplate anything but just paying it.[31] There's also a three-year exemption for new facilities, more exemptions based on R&D investments, and other offsets based on rules that still need to be negotiated.

Minister Baird tried to sell the plan as a balanced approach between the "perfection" demanded from environmentalists and the status quo preferred by many in the industry. The reality is that the new rules—especially the intensity-based approach, the 2010 start date, the three-year grace period for new projects and the fifteen-dollar escape clause—seem to be carefully crafted to have little effect on oil production from the tar sands, which are expected to triple by 2015.

The End Game?

It is impossible to predict what will happen in federal climate-change policies in Canada over the coming months and years. At the time of this writing, there is a bill in the Senate that will compel the government to come up with a plan to meet Canada's Kyoto targets. If the *Kyoto Implementation Act* is passed, as expected, it remains to be seen how the government and the opposition will react.

Other than this potential bill or others coming from the opposition, there is little chance of advancing climate policies that will provoke action in the immediate future. The heavily amended *Clean Air and Climate Change Act*, as announced by the Conservative House leader, will not be brought back for second and third reading.

Meanwhile, the government's industry regulations will be finalized over the next three years. With a stipulation that these be reviewed every five years, you can expect industry associations—such as the Canadian Association of Petroleum Producers and the Canadian Manufacturers and Exporters—to work hard to make sure the details favour their members in the short term and get weaker when reconsidered.

Internationally, United Nations negotiations will continue to discuss and, hopefully soon, decide what the next phase of the Kyoto Protocol will look like. All indications from the federal government are that little effort will be put into these. Instead Prime Minister Harper will likely try to lend momentum to smaller US-led coalitions on climate change that are voluntary, ineffective and have, in some instances, undermined the UN negotiations. The Asia-Pacific Partnership on Clean Development and Climate is a typical initiative, a voluntary agreement between six countries—the US, Australia, Japan, China, India and South Korea—whose own documents show a doubling of GHG emissions from those countries would be a best-case scenario.[32] Last year, Minister Ambrose indicated that the approach of the partnership is "very much in line with where our government wants to go."[33]

So despite climate change becoming a top of mind issue in Canada, it still looks like real and—according to the best scientific and economic analysis—necessary action is still not imminent. But expect the politics and the numbers games to continue.

Notes

1. Nicholas Stern, "Stern Review on the Economics of Climate Change," 2006, http://www.hm-treasury.gov.uk/independent_reviews/stern_review_economics_climate_change/stern-review_index.cfm.
2. Commissioner of the Environment and Sustainable Development, "2006 Report of the Commissioner of the Environment and Sustainable Development," http://www.oag-bvg.gc.ca/domino/reports.nsf/html/c2006menu_e.html.
3. The Intergovernmental Panel on Climate Change (IPCC) is a scientific organization comprising over 2000 of the world's leading scientists on climate change. The IPCC's three working groups released their conclusions in stages between February and May 2007. See Intergovernmental Panel on Climate Change, "Working Group 1: The Physical Basis of Climate Change," 2007, http://ipcc-wg1.ucar.edu/wg1/wg1-report.html.

4. For example, in the summer of 1998, Canada hosted the Toronto Conference on the Changing Atmosphere, a gathering of climate scientists that called for 20 percent reductions in global GHG emissions by the year 2000.

5. For example, economic studies on climate change mitigation have been undertaken by Sweden, Norway, Germany, Canada and California—all of which showed no or little economic cost to reducing emissions. Canada's National Climate Change Process had fourteen roundtables of experts that investigated the technologies and policies required to reduce greenhouse gases from different sectors. The process was concluded with economic modeling that showed that Canada could reach its Kyoto commitments at a cost of less than 1 percent of GDP over ten years (the economy would expand by 29 percent instead of 30 percent over the course of one decade). See Analysis and Modeling Group, "An Assessment of the Economic and Environmental Implications for Canada of the Kyoto Protocol," National Climate Change Process, 2000.

6. For an example of Canadian public concern on climate change, see Brian Laghi, "Climate concerns now top security and health," *Globe and Mail*, January 26, 2007.

7. The documents were never obtained publicly. Media reports include Bill Curry, "Ottawa now wants Kyoto deal scrapped," *Globe and Mail*, May 19, 2006.

8. CBC, "Ambrose slams Liberals at UN climate summit," November 15, 2006, http://www.cbc.ca/world/story/2006/11/15/ambrose-summit.html.

9. See Louis-Gilles Francoeur, "Le Canada perd la face à Bonn," *Le Devoir*, May 16, 2006 and François Cardinal, "Berlin rabroue le Canada pour son défaitisme," *La Presse*, May 16, 2006.

10. Mike De Souza, "Canada criticized over Kyoto rejection," Canwest News Service, November 16, 2006.

11. Alan Findlay and Vivian Song, "Have we no shame?" *Toronto Sun*, March 1, 2007.

12. Leahy, Stephen. 2006. "Canada Reneges on Kyoto Climate Change agreement." *Word Press*. November 29, 2006.

13. Environment Canada. (undated). "Memorandum to Minister: Clean Air, GHG, and Congestion Impacts of Tax Incentive for Transit Riders." Leaked document.

14. Government of Canada. 2006a. "The Budget Plan 2006: Focusing on Priorities."

15. Government of Canada. 2006b. "Clean Air Act." Available at: http://www2.parl.gc.ca/HousePublications/Publication.aspx?Docid=2413797&file=4.

16. Government of Canada. 2007a. "Clean Air and Climate Change Act." Available at: http://www2.parl.gc.ca/HousePublications/Publication.aspx?Docid=2826031&file=4.

17. Canada, Parliament of Canada, Official transcript of 39th Parliament, 1st Session, Legislative Committee on Bill C-30, March 29, 2007.

18. CBC, *The House*, March 30, 2007 (radio interview).

19. Bea Vongdouangchanh, "Tories have 'no intention' of passing Clean Air Act: Van Loan," *The Hill Times*, April 30, 2007.

20. David Boyd, "Sustainability Within a Generation: A New Vision for Canada," prepared for the David Suzuki Foundation, 2004.

21. Matthew Bramley, "The Case for Deep Reductions: Canada's Role in Preventing Dangerous Climate Change" (David Suzuki Foundation and the Pembina Institute, 2005).

22. IPCC, "Working Group 1."

23. David Suzuki Foundation, "Intensity-based Targets: Not the Solution to Climate Change," briefing note, 2007.

24. Environment Canada, "Canada's Greenhouse Gas Inventory 2004," 2006.

25. Government of Canada, "Project Green: Moving Forward on Climate Change," 2005.

26. Natural Resources Canada, "Climate Change: Large Final Emitters Strategy," documents

obtained through Access to Information, 2003.

27. David Suzuki Foundation et al., "Assessment by Canadian Environmental Leaders of the Government's Kyoto Implementation Plan," press release, 2005, http://www.davidsuzuki.org/files/climate/KyotoLeadersStatement.pdf.

28. Government of Canada, "The Budget Plan 2007: Aspire to a Stronger, Safer, Better Canada," 2007.

29. Government of Canada, "The Cost of Bill C-288 to Canadian Families and Businesses," 2007.

30. Government of Canada, "Regulatory Framework for Air Emissions," 2007.

31. Ibid.

32. Climate Institute, "First Progress Report on the Asia-Pacific Partnership on Clean Development and Climate," 2006.

33. CTV, "Tories consider U.S.-led effort to fight pollution," April 25, 2006, http://www.ctv.ca/servlet/ArticleNews/story/CTVNews/20060425/emissions_060425/20060425?hub=CTVNewsAt11.

A Twelve-Step Program to Combat Climate Change

by Cy Gonick and Brendan Haley

> "Climate change is not just a moral question: it is the moral question of the 21st century. There is one position even more morally culpable than denial. That is to accept that it's happening and that its results will be catastrophic; but fail to take the measures needed to prevent it."—George Monbiot, *Heat: How to Stop the Planet from Burning.*

Global warming is now garnering citizen's attention around the world. In Canada, the Harper government's abandonment of climate policy has awakened the public to the need for action. In October 2006, Harper attempted to hoodwink us by utilizing a public-relations strategy taken straight from George Bush—namely, promise "clean air" and phony targets for emissions that mirror business as usual, while raising doubt about the science of global warming and the economic consequences of taking action.

Canadians saw through the Harper government's ruse. So in the spring of 2007, Harper unleashed a new, much more deceptive public-relations campaign after replacing Rona Ambrose with bully-boy John Baird. The best this chastised bunch could come up with is a selection of the pitifully inadequate climate change programs earlier enunciated by Stéphane Dion when he held the environment portfolio under the former Liberal government, and yet more inadequate targets and phony "intensity-based" regulations for large industries.

A Looming Catastrophe

Climate change is almost certainly the greatest threat to life on earth. In economic terms, according to the Stern Report (Nicholas Stern, formerly chief economist of the World Bank), left to itself global warming will cause a massive drop in annual global gross domestic product of up to 20 percent by the end of this century. This would produce a crisis on the scale of the Great Depression. Unlike that downturn, this one would be permanent, not cyclical.

Long before then, however, a combination of floods, drought and famine will have destroyed many countries in the poorer parts of the earth. Even now, with average temperatures only 0.7°C higher than norms in the pre-industrial era, just about every country in the world has experienced weather-related crises of one kind or another—record heat waves in 2006 in the US, Brazil, Europe; a drought in Africa followed by the worst flooding in recent history; heavy rainfall in the Sahara desert causing the displacement of 600,000 people; 200,000 forced to evacuate in New England and the northeastern US; record rainfall in Vancouver; 1000 people dead in China from the worst storms in a decade; over 500 lives lost in the Philippines and many more still missing in a typhoon that affected a million and a half people.

With arctic ice declining by 8.6 percent a year, the Arctic Ocean could be open water by 2040 if not earlier. Permafrost thaw is reported in northern Canada, Siberia, most Arctic land, the Alps and Tibet, possibly releasing billions of tonnes of methane into the atmosphere. Methane gas is about twenty times more potent a greenhouse gas than carbon dioxide.

As average global temperatures approach 5°C above historical levels—which scientists say will happen if we continue with business as usual—it would mean the end of the world's greatest cities and cause human migration on a scale never before experienced.

Two Degrees and Risin'

The scientific consensus is that 2°C above pre-industrial global temperature levels is the threshold of danger. The potential for large and potentially irreversible shifts in the climate system can occur once we reach 2°C and beyond. The probabilities show that we may have already passed this level.

Concentrations of global warming pollution in the atmosphere have to stabilize at today's levels, or only slightly higher. If global emissions peak by 2015 and quickly decline we have a chance of avoiding 2°C.[1]

As the *Guardian* editorialized on August 11, 2006, "If we don't take action soon, we could unleash runaway global warming that will be beyond our control and it will lead to social, economic and environmental devastation worldwide. There's still time to take action, but not much."

If it is to be successful, any workable pollution reduction must consider issues of equity, both globally and within nations. Pollution reductions cannot be achieved by keeping the population of the global south in poverty.

The North has both an historic responsibility as well as the capacity to reduce emissions. In addition, it is hardly fair that the most severe impacts will be felt by regions that have contributed least to the problem through the burning of fossil fuels. It is also the poor within industrialized nations that will be impacted most severely, as the worst effects of New Orleans' Hurricane Katrina so clearly demonstrated.

A Radical Transformation

A few calculations show how radical a transformation will be necessary to achieve an equitable reduction path. Let us assume that the biosphere is currently able to absorb about half of the 7.9 billion tonnes of carbon emitted per year.[2] If the ability to emit the remaining carbon is divided equally amongst the world's 6.5 billion inhabitants, each person could emit 0.6 tonnes of carbon per year. Currently, the Canadian average is about 6.2 tonnes of carbon per person, if we consider all emissions.[3] Thus about a 90 percent reduction from current levels is in order for Canada.[4]

If the biosphere's ability to act as a sink for carbon reduces, the global population increases or the effects of warming gases become more severe (all are likely) the reductions needed become more stringent. The National Round Table on the Environment and Economy is calling for current emissions to be reduced by 60 percent by 2050. If we consider global equity, a reduction of more than 90 percent from current levels will be needed by that time, if not before.

Massive changes in the structure of the economy and in the personal consumption patterns of all of us are essential if we are to achieve the necessary reductions of greenhouse gases in the atmosphere. Can this be done by soliciting changes in consumption and investment decisions through adjustments to our tax system and by creating a market for carbon along with capping emissions for industry? This is the policy model recommended by mainstream economics and endorsed to some degree or another by governments in Europe and the US. It is also the basis of the policies being put forward by all of Canada's political parties, Greens and NDP included. They differ only in the details and in the sense of urgency.

When Canada and later the US entered World War II, they both had to convert their industries to a war-time footing. Canada was required to build some crucial industries from scratch. In this life-or-death situation neither relied on market mechanisms to achieve such large changes in so swift a manner. They adopted an economic planning model, directly reallocating capital to war industries, regulating industry, rationing resources, instituting price controls and, in Canada's case, placing much essential industry under government ownership and/or control.

The evidence is overwhelming that with climate change we are facing another life-or-death situation. This time, however, victory can be ours only with a permanent change in the way we organize our economic lives. We are not advancing visions of austerity. We can have modest economic growth and decent living standards for all so long as we develop the economy in harmony with the environment. Nor are we saying that there is no role for the market in bringing about the necessary changes in economic behaviour.

Merely that it must play a subordinate role.

In developing a reform plan to deal with climate change, the labour movement and environmentalists ought to recognize that ending the exploitation of both workers and the environment creates new opportunities for radical politics. The environmental movement must recognize that ecological politics is profoundly integrated with the national and international distribution of wealth and power, that the destruction of the environment is intimately linked with the process of capital accumulation and that labour is a fundamental player in transforming the nature of industrial production. Labour must accept socially responsible production as a new front and recognize how the call for environmentally friendly production can re-invigorate the struggle for democratic workplaces and better jobs.

Any climate-change policy must include "just transitions" for workers. This is not a defensive policy, but a catalyst to the green transition. Workers must become enthusiastic contributors to environmentally friendly production and innovation and see opportunities to transition towards better jobs.

Here are some suggestions for further discussion. They are taken from a variety of sources and are not restricted to the federal level of government.

- Redesign our cities so that more people travel much less. Make it possible for them to live, work and shop in the same walkable neighbourhood with travel engaged in less for commuting and more for pleasure.
- Upgrade our public transit system with more affordable fares and vastly improved service so that it can compete with automobiles on a level playing field. Finance the upgrade through green municipal investment funds and funds shifted from road building and road widening programs.
- Develop an individual or family-based rationing system for selected carbon intense activities such as airline travel or even car travel, rationing being a more democratic way of restricting consumption than taxation.
- Require all new single family and multi-family housing to meet net-zero energy standards—better insulated and sealed homes with higher quality windows and doors, high-efficiency furnaces and solar heating and power systems.
- Ban the sale of all lighting, appliances and equipment that do not meet Energy Star standards.
- Introduce mandatory energy audits and other efficiency programs for homes, businesses, government and community facilities. Provide insulation upgrades, air sealing, motor replacement and refrigeration improvements and the like—to be financed by a combination of federal subsidies and user-fees on utility bills. Oblige landlords and homeowners to meet high energy-efficiency standards before resale of buildings or units.
- Undertake a comprehensive inventory of all industrial processes (in col-

laboration with unions) focussed on reducing waste through processes such as combined heat and power generation, upgrading production processes and shifting production lines where necessary.

- Mandate fuel-efficient and GHG standards for automobiles and trucks that are on par with Europe and Japan (more stringent than California standards).
- Legislate an immediate moratorium on the development of the tar sands, which would otherwise contribute half of all emissions growth over the next twenty years. The moratorium would continue at least until the industry has developed widely proven technology for carbon storage with permanent sequestration. A similar moratorium would be imposed on new coal mines until carbon storage and sequestration are assured.
- Legislate an immediate end to the $1.4 billion annual subsidies to the oil and gas sector.
- Introduce absolute caps on industrial emissions with the caps being progressively reduced. Firms that exceed these emission caps would be subject to heavy fines and shut down or taken into public ownership if their negligence persists.
- Strategically upgrade Canada's electricity grid for sustainable energy, increasing east-west connections and promote offshore wind and biomass fuels.

We regard these twelve steps as examples of the kind of an action program that must be instituted very soon. They will not stop global warming. They are designed to slow it down, thereby limiting the devastation that would otherwise occur. Beyond them must lie an eco-socialist future that will fundamentally change the way we organize ourselves, the way we produce and what we produce and the way society's limited resources are used to meet the needs of all its members in an equitable and democratic way.

Notes

1. See the Working Group III contribution to the Intergovernmental Panel on Climate Change Fourth Assessment Report, *Climate Change 2007: Mitigation of Climate Change Summary for Policymakers.*
2. "The world currently emits 7.9 billion tonnes and about half are sequestered." Quoted on the Commonwealth Scientific and Industrial Research Organisation web page, "Increase in carbon dioxide emissions accelerating," http://www.csiro.au/news/ps2im.html.
3. 2006 Environment Canada Inventory has 758 Mt of carbon dioxide emitted; CIA *World Factbook* sets Canadian population at 33.39 million. This means 22.7 tonnes of carbon dioxide per person. To convert from carbon dioxide to carbon, divide by 3.666666666. This calculates as 6.19 tonnes of carbon per person. Carbon Dioxide Information Analysis Center, http://cdiac.ornl.gov/trends/emis_mon/emis_mon_co2.html.
4. The calculation is (6.19-0.6)/6.19 = 0.90.

Statistical Appendix on World Oil Reserves

by John W. Warnock

World Oil Reserves by Country as of January 1, 2006[1]

Country	Oil Reserves (billion barrels)
Saudi Arabia	264.3
Canada	178.8
Iran	132.5
Iraq	115.0
Kuwait	101.5
UAE	97.8
Venezuela	79.9
Russia	60.0
Libya	39.1
Nigeria	35.9
United States	21.4
China	18.3
Qatar	15.2
Mexico	12.9
Algeria	11.4
Brazil	11.2
Kazakhstan	9.0
Norway	7.7
Azerbaijan	7.0
India	5.8
Rest of World	68.1
World Total	1,292.5

The above table is for "proven reserves," oil that can be extracted using present technology and given the economics of the time. Liberal economists argue that reserves expand with new technology and the rise in the price of oil.

IHS Energy, based in Geneva, has the most extensive database in the world on oil and gas production, and their findings are used by most private corporations and government agencies. Their recent reports stress that the average size of new discoveries has been steadily declining since 1975, the industry has increasingly depended on deep-water discoveries in recent years as on-shore discoveries decline, and that since the early 1980s the rate of oil extraction has exceeded reserves replacement.

There are some controversies around the figures in the above table. First, the reserves for the OPEC countries are questionable. Data are closely guarded by the OPEC governments. The quotas set by OPEC for each country's share of production are based on their share of total OPEC reserves. Over the 1987 to 1988 period, many of the OPEC governments reported dramatic increases in their reserves, apparently in an attempt to increase their production quota.

Canada's conventional oil reserves, including heavy oil, are estimated at between 2.9 billion barrels[2] and 4.4 billion barrels.[3] British Petroleum, in its widely cited *Statistical Review of World Energy*, lists Canada's oil reserves at 16.8 billion barrels. They include present operations in Alberta's tar sands. The above figure from the *Oil and Gas Journal* includes an estimate of the oil that can be extracted the bitumen of the Alberta tar sands using present mining and in situ technology.

The reserve figures for Venezuela include 45 billion barrels of convention oil and 35 billion barrels of extra heavy oil being extracted from four zones in the Orinoco oil belt. It is estimated that Venezuela has 1.2 trillion barrels of extra heavy oil of which 270 billion barrels can be extracted. The heavy-oil reserves are not only larger than Canada's tar sands, they are being extracted by cold heavy-oil production with sand (CHOPS), a process that uses 30 percent cold water in an emulsification. These heavy-oil wells produce around 850 barrels per day, and in 2006 the total lifting cost was less than one dollar per barrel. Extraction without using heat, better fields and easier pipeline transportation make extraction much more profitable than the Alberta tar sands. These reserves are not classified by OPEC as proven reserves as this would greatly upset the system of setting production quotas.

Notes

1. Source: "Worldwide Look at Reserves and Production," *Oil and Gas Journal* 103, 47 (December 19, 2005): 24–25. The United States Department of Energy and other agencies accept this industry estimate, which is almost identical to that set by the US Geological Survey.
2. Statistics Canada, Human Activity and the Environment. Annual Statistics, 2005, Catalogue No. 16-201-XIE, 2006.
3. Canadian Association of Petroleum Producers, "2004 Petroleum Reserves Estimate," November 2005, http://www.capp.ca.